Problem Solving in

CONCEPTUAL
Physics

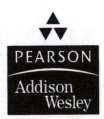

Paul G. Hewitt

City College of San Francisco

Phillip R. Wolf

Mt. San Antonio College

PEARSON

Addison
Wesley

San Francisco   Boston   New York
Cape Town   Hong Kong   London   Madrid   Mexico City
Montreal   Munich   Paris   Singapore   Sydney   Tokyo   Toronto

Editor-in-Chief: Adam Black
Senior Acquisitions Editor: Lothlórien Homet
Project Editor: Liana Allday
Senior Production Supervisor: Corinne Benson
Main Text Cover Designer: Yvo Riezebos Design
Supplement Cover Manager: Paul Gourhan
Supplement Cover Designer: Joanne Alexandris
Senior Manufacturing Buyer: Michael Early
Cover and Text Printer: Banta Book Group
Cover Credits: **Wave and surfer** Photolibrary.com/AMANA AMERICA INC. IMA USA INC.;
particle tracks Lawrence Berkeley National Laboratory, University of California.

ISBN   0-8053-9377-3

Problem Solving in Conceptual Physics
Table of Contents

To The Student

Concepts first; calculations later. This has always been the credo of *Conceptual Physics*. The intent of *this* book is to extend your present understanding of physics concepts. Problems can be a way to appreciate the connections between concepts. Whereas many physics problems are more math than physics, you'll find the opposite in this book. Number crunching is not the name of the game here. In fact, numbers are not given in the main part of the problems. Rather, problems are phrased in terms of mass m, speed v, force F, and so on, putting the focus on the concepts themselves. In a second part of most problems, numbers are given so that you can transform a concept solution to a numerical one. Dealing with letters rather than numbers may take a bit of getting used to, but you may soon discover the exhilaration that comes from being able to learn a thing well. You may also recognize that a wide variety of seemingly different problems are variations on the same root concept, which invite similar solutions. (This is thinking like a physicist!)

Generally, the problems in each chapter progress from simpler ones to more challenging ones, with a wide assortment to choose from.

Enjoy!

Paul G. Hewitt

Phillip R. Wolf

To The Instructor

For the past fifty or more years, problem solving is what physics courses have been all about—something with which I have been at odds. I think we can all agree that techniques of problem solving are very important to a physics major. But what irks me is problem-solving becoming an end-all to an introductory physics course. Because of the time involved with problems, instruction that should be instilling a feeling for concepts is relegated to the back seat. Undue emphasis is given to teaching the tools of physics rather than the physics itself. Students end up memorizing recipes and math techniques to solve problems that give them the false impression that they're learning physics. I wrote *Conceptual Physics* to combat this widely-accepted practice. Because of its focus on concepts, with problem solving secondary, *Conceptual Physics* is widely adopted—but mainly for classes of non-science students.

Many instructors have expressed interest in a problems supplement to *Conceptual Physics* so that it could be the text for an algebra-trig course—hence this problems supplement. The selection of problems herein emphasizes the physics rather than the algebra and trig. Phil and I have avoided puzzles which stress mathematical manipulation and are more math than physics. Problems herein are meant to enlighten your average students, rather than just to challenge your best and brightest. Some are challenging, and are so labeled with a bullet (•). Physics more than math is the thrust of this book.

It's often said that the number of problem-solving strategies is about equal to the number of instructors teaching them. In my algebra and calculus based courses I've tried several strategies over the years and reduced them to the simple method illustrated on the following sample solutions. Knowing how to get started is a primary difficulty for students. We address this by first focusing on what is being asked for. For example, if the problem asks for speed, the first step is writing "$v = ?$" I give my students 40% credit if they begin a solution this way. Then identify the physics concept that underlies the problem. If the problem involves collisions, then consider conservation of momentum. If the problem concerns forces and motion, then consider the work-energy theorem, etc. (This weans students away from the practice of categorizing problems as pulley problems, inclined plane problems, and so on.) Then write that relationship in equation form. Unless your equation directs otherwise, avoid searching for a perceived missing mass, acceleration, or whatever when it's not needed. When the equation has more than one term that represents unknown quantities, the procedure is repeated. Steps are guided by the equation.

Some conceptual physics instructors downplay equations and discourage the practice of grabbing for an equation when confronted with a problem. I take the opposite view. I think grabbing for equations should be encouraged—especially when equations are seen as abbreviated statements about the way variables connect and relate to one another. The terms in an equation are akin to notes on a musical scale. Just as notes guide fingers on a keyboard, equations can guide thinking. I teach my students to see central physics relationships in equation form, favoring those that state first principles: Newton's second law, conservation of momentum, conservation of energy, work-energy theorem, Newton's law of gravitation, Boyle's law, Ohm's law, Faraday's law, and so on. Equations can guide a student's thinking.

You'll note that sample solutions in this book follow this procedure, focusing first via $v = ?$, $t = ?$, $F = ?$, and so on. Then central principles are stated in equation form. From the equation the term asked for is isolated. You'll note some exceptions to strictness in significant figures here and there to avoid unnecessary digits. We'll say that 100 cm = 1 m, for example. And we'll say that $2 \times 6 = 12$, and not the strictly correct $2 \times 6 = 10$. Simple and effective.

Chapters 1 and 2 don't lend themselves to problems, so we begin with Chapter 3: Linear Motion. This is the only chapter with no physical laws. But it has definitions that lay the foundation for the laws of the following chapters.

Let's begin!

PGH

To The Instructor

What is important for students to learn? How do we get them there? In my teaching with *Conceptual Physics* I tell my students that although they won't be doing a lot of math in the course, they'll be doing a LOT of mathematical thinking. There is a world (at least!) of physics embodied in equations such as $a = F_{net}/m$, and a ton of conceptual understanding that can occur without dealing with numbers and solving physics problems. Yet well-chosen physics problems *can* lead students to a deeper understanding of the physics, with the equations serving to guide their thinking.

This book provides good problems that students in a typical *Conceptual Physics* course can solve. We gently introduce trigonometry and topics such as springs and coefficients of friction to expand upon material touched on in *Conceptual Physics* and traditionally covered in a problem-solving course. Our hope is that you and your students can have the best of *both* worlds—the richness of a concept-focused approach to physics *and* the clarification of understanding that comes from using concepts to solve problems.

PRW

Detailed solutions to all problems are on the website in the Instructor's Resource Area.

Acknowledgements

We gratefully acknowledge the intensive input of Ken Ford, Herb Gottlieb, David Housden, and Diane Riendeau. For valued suggestions and feedback we are grateful to Tsing Bardin, Howie Brand, George Curtis, Marshall Ellenstein, Jim Hicks, Chelcie Liu, Fred Myers, Stan Schiocchio, and David Williamson. Thanks go to Jesse David and Ellender Wall for adaptations of several original and insightful problems from their book *Introductory Physics—A Problem-Solving Approach* (Analog Press, 1997). We're also grateful to young Gretchen Hewitt Rojas for keyboarding help, and to Ernie Brown for designing the chapter headers and the physics logo. The person who has worked as faithfully on the book as the authors is Paul's wife Lillian Lee Hewitt, to whom we are enormously grateful. For their support and encouragement we thank Phil's wife Mala Arthur and son Zephram Wolf. Thanks also to the many students at Mt. San Antonio College who provided valued feedback.

3 Linear Motion

Kinematics is about *describing* motion. In kinematics there are no laws of physics—only concepts and definitions—*distance, displacement, speed, velocity,* and *acceleration,* and the relationships among these concepts. What's especially important in kinematics is to have clear the distinction between velocity and acceleration so that Newton's second law of motion can be better understood in subsequent chapters.

Distance as used in physics is the same as distance in everyday usage—how far something moves. It's measured in meters (or some other length unit).

Displacement is a little trickier. It's the straight-line change in position between the starting and ending points of motion. It's a *vector* quantity since it involves both a distance and a direction. After completing one circuit on a racetrack, for example, a car has *zero* displacement, since it ends where it started (its change in position is zero). For motion in a straight line (linear motion) in a single direction, there is no need to distinguish between distance and displacement. They are, for practical purposes, the same. In this chapter, we shall simply use distance.

Speed, $v = \dfrac{\text{distance}}{\text{time}} = \dfrac{d}{t}$, is the first kinematics equation we study.

Average speed $= \dfrac{\text{total distance covered}}{\text{time interval}}$; $\bar{v} = \dfrac{d}{t}$.

The bar over the symbol v indicates "average" speed. Rearrangement gives,

$d = \bar{v}\,t$. That is, distance covered equals average speed × time.

Velocity is expressed with the same equation, but involves direction. Velocity is a vector quantity. In problems involving velocity, direction is either implied or specified. Combining velocities must be done by vector rules, which are treated in Chapter 5. The standard unit for measuring speed and velocity is meters per second (m/s).

When the velocity (or speed) changes at a steady rate, average velocity (or average speed) is the average of the initial and final speeds or velocities—their sum is divided by 2. That is,

$\bar{v} = \dfrac{v_0 + v_f}{2}$, where v_0 is the *initial* velocity, and v_f is the *final* velocity.

Distance traveled is found by multiplying the average speed by the time interval. (Average velocity multiplied by time gives *displacement,* a vector quantity that we won't make a big deal of.)

$d = \dfrac{v_0 + v_f}{2}t$.

Acceleration is the rate at which velocity changes:

$$a = \frac{\text{change in velocity}}{\text{time interval}} = \frac{\Delta v}{\Delta t} = \frac{v_f - v_0}{\Delta t}.$$

Along a straight-line path, acceleration can be defined as the change in *speed* per time interval. Then the same equation applies to both speed and velocity. In this chapter we consider only uniform acceleration, motion that changes at a steady rate. The standard unit of acceleration is meters per second per second, or meters per second squared (m/s^2).[*]

Simple rearrangement of $a = \dfrac{\Delta v}{\Delta t}$ gives $v_f - v_0 = at$, which says the change in velocity = the acceleration $\times$ time.[•] Further rearrangement produces

$$v_f = v_0 + at.$$

This equation says that the velocity of an object at a time t (instantaneous velocity) equals its initial velocity plus at, the additional velocity acquired due to acceleration during this time.

> When the initial velocity is zero (object starting from rest) the final velocity of the moving object simply equals the velocity acquired during acceleration,
>
> $$v_f = at.$$
>
> For free-fall cases where an object falls from rest and air resistance is negligible, velocity acquired is given by the equation
>
> $$v_f = gt.$$
>
> where g is the acceleration due to gravity (10 m/s^2 for estimation and 9.8 m/s^2 for calculation).

The equations above are not laws of physics, but are simply definitions and relationships expressed in mathematical notation.

> Distance covered during accelerated motion is given by
>
> $$d = v_0 t + \tfrac{1}{2}at^2.$$

(The derivation of this equation is in Appendix B in your textbook.)

> Distance covered during free fall is then,
>
> $$d = v_0 t + \tfrac{1}{2}gt^2.$$

---

[*] For example, an acceleration of $2\frac{m}{s^2} = \frac{2\frac{m}{s}}{1 s}$ means that every second the speed (or velocity) is increasing by ?

[•] For example, an acceleration of $2\frac{m}{s^2}$ for 3 seconds gives a change in velocity of $6\frac{m}{s}$. That is, $2\frac{m}{s^2} \times 3s = 6$

If you designate the direction up as + and down as –, then the acceleration of the object is $-g$ and

$$d = v_0 t - \tfrac{1}{2} g t^2.$$

For objects falling from rest, downward distance is simply $d = \tfrac{1}{2} g t^2$.
(If we use $g = 10$ m/s^2, this has the magnitude $d = 5t^2$. For $g = 9.8$ m/s^2, we get $d = 4.9\, t^2$.)
Notice that we express units of measurement in standard font and the variables representing measured quantities in italics.

Acceleration is uniform when the force that produces it is constant. The force due to gravity at Earth's surface, for example, is constant on any given object, so g (the acceleration due to gravity) is constant. But the acceleration of an object oscillating on the end of a spring, for example, is not uniform. That's because the force from a stretched spring varies as the spring is stretched (springs are treated in Chapter 7). We consider only uniform acceleration here in Chapter 3.

A bit of mathematical manipulation with the above equations produces a formula that does not include time (obtained by solving one equation for time and plugging it into another).[*] The formula is

$$v_f^2 - v_0^2 = 2ad. \text{ This can be rearranged to read}$$

$$d = \frac{v_f^2 - v_0^2}{2a}. \text{ Or with further rearrangement, } a = \frac{v_f^2 - v_0^2}{2d}.$$

These three formulas are useful in many problem-solving situations—especially those where time is not given.

Be aware that acceleration is one of the most misunderstood concepts in physics! It is often confused with velocity, but is fundamentally different. When answering questions about acceleration, think in terms of the equations for acceleration. Solving the problems that follow will help.

Because there's so much good physics to cover in your course, your instructor may hurry you through this chapter. Don't fret, for the concepts of kinematics are employed in following chapters where you'll have opportunities to develop a deeper understanding of them. It would be a shame to get bogged down in this part of the course, which is devoid of physical laws. Grasp the ideas of speed, velocity, and acceleration, and then move on to where they're useful!

On the next page is a summary of useful linear motion equations (in which acceleration a is a constant).

---

[*] From $d = \bar{v} t$ and $t = \dfrac{v_f - v_0}{a}$ $\Rightarrow d = \left(\dfrac{v_f + v_0}{2} \right)\left(\dfrac{v_f - v_0}{a} \right) = \dfrac{v_f^2 - v_0^2}{2a}.$

$$\bar{v} = \frac{d}{t} \quad \Rightarrow d = \bar{v}t = \left(\frac{v_0 + v_f}{2}\right)t$$

$$a = \frac{\Delta v}{\Delta t} = \frac{v_f - v_0}{t} \quad \Rightarrow \quad v_f = v_0 + at$$

$$d = v_0 t + \tfrac{1}{2}at^2$$

$$v_f^2 - v_0^2 = 2ad$$

Most problems in this book begin by asking for a generalized solution, expressed in symbols. This is usually followed by a step that asks for a numerical solution together with appropriate units of measurement. Let's consider some sample problems and their solutions:

Sample Problem 1

While driving along the highway at constant speed v you sneeze, and your eyes close for a brief time t.
(a) How far along the highway do you travel during your sneeze?

> *Focus*: $d = ?$ (We focus on what is being asked for, in this case distance traveled.)
>
> We start with $v = \dfrac{d}{t}$ (beginning with a basic definition, in this case the equation that defines speed, which includes the distance we're looking for.) Rearranging gives $d = vt$, so the distance traveled with closed eyes is your average speed multiplied by the time of the sneeze.

> *Solution*: The answer is $d = vt$.

(b) Calculate the distance (in meters) that you travel while sneezing given that your highway speed is 115 km/h and your eyes close for 0.70 s while sneezing.

> *Focus*: Here we're asked to find the distance d in meters, while speed is given in km/h. Our task is to convert 115 km/h to m/s. We use a process called "units analysis." We set up conversion factors so the km cancel, and hours cancel leaving us with m/s. The conversion factors we need are: 1 hr = 3600 s, and 1000 m = 1 km.
>
> $$115\frac{km}{h} \times \left(\frac{1h}{3600s}\right) \times \left(\frac{1000m}{1km}\right) = 31.9\frac{m}{s}$$

> *Solution*: Now, $d = vt = \left(31.9\frac{m}{s}\right)(0.70s) = \mathbf{22\,m}$.
> Again, we display the answer in **bold**.

Summary for converting units
- Start with the quantity to be converted. In this case, 115 km/h.
- Multiply by a ratio (called a *conversion factor*) that's equal to 1; the quantity on the top is equal to the quantity on the bottom, but expressed in different units. In this case, the conversion factors are $\left(\dfrac{1000 \text{ m}}{1 \text{ km}}\right)$, and $\left(\dfrac{1 \text{ h}}{3600 \text{ s}}\right)$.
- The ratio is arranged so that multiplying and canceling gets rid of the unwanted units and leaves units that you want.

The problem asks for distance in meters. But suppose it had instead asked for the distance in feet. We know there are 3600 seconds in 1 hour, 1000 m in 1 km, 100 cm in 1 meter, 2.54 cm in 1 inch, and 12 inches in 1 foot. Then

$$115\frac{\text{km}}{\text{h}} \times \left(\frac{1\,\text{h}}{3600\,\text{s}}\right) \times \left(\frac{1000\ \text{m}}{1\ \text{km}}\right) \times \left(\frac{100\ \text{cm}}{1\ \text{m}}\right) \times \left(\frac{1\ \text{in}}{2.54\ \text{cm}}\right) \times \left(\frac{1\ \text{ft}}{12\ \text{in}}\right) = 105\,\frac{\text{ft}}{\text{s}}.$$

We see that the units themselves guide the mathematical operation. There are many ways to do the same conversion. We could have used 60 seconds = 1 minute and 60 minutes = 1 hour if we didn't realize that 3600 seconds = 1 hour. We could have used 3.28 ft = 1 m if we didn't use 2.54 cm = 1 in. Units guide the process.

Sample Problem 2

Mala enjoys exercise and jogs a distance x at a constant speed v.
(a) What time does it take Mala to cover distance x?

Focus: $t = ?$

From $v = \dfrac{d}{t} \implies t = \dfrac{d}{v}.$

Solution: $t = \dfrac{d}{v} = \dfrac{x}{v}.$

(b) Calculate the time in minutes for Mala to jog 1.3 km at a constant speed of 2.0 m/s.

First convert kilometers to meters:

$$1.3\,\text{km} \times \left(\frac{1000\,\text{m}}{1\,\text{km}}\right) = 1300\,\text{m}. \text{ Then } t = \frac{x}{v} = \frac{1300\ \text{m}}{\left(2.0\frac{\text{m}}{\text{s}}\right)} = 650\,\text{s}.$$

Solution: By units analysis, $650\ \text{s} \times \left(\dfrac{1\ \text{min}}{60\ \text{s}}\right) = 11\,\text{min}.$

Sample Problem 3

A bus starting from rest uniformly accelerates along a level road.
(a) How far does the bus travel when accelerating from rest to speed v_f in a time interval t?

Focus: $d = ?$

From $\bar{v} = \dfrac{d}{t}$, $d = \bar{v}t = \left(\dfrac{v_0 + v_f}{2}\right)t$, and since v_0 is zero, $d = \dfrac{v_f t}{2}.$

(We get the same answer if we use $d = v_0 t + \frac{1}{2}at^2$, where $v_0 = 0$ and $a = \frac{\Delta v}{\Delta t} = \frac{v_f}{t}$. Then

$$d = v_0 t + \tfrac{1}{2}at^2 = 0 + \tfrac{1}{2}\left(\frac{v_f}{t}\right)t^2 = \frac{v_f t}{2}.)$$

(b) Calculate the distance traveled by the bus when it starts from rest and accelerates uniformly to 12 m/s in a time of 5 s.

Solution as above: $d = \dfrac{v_f t}{2} = \dfrac{\left(12\frac{\text{m}}{\text{s}}\right)(5\text{s})}{2} = 30\ \text{m}.$

Sample Problem 4

Two bicyclists ride toward each other on a long unobstructed straight road, each riding at a constant speed v. When the bikes are distance x apart, a bee begins flying from the front wheel of one bike to the front wheel of the other bike at a steady average speed of $3v$. When the bee reaches each wheel it abruptly turns around and flies back to touch the other wheel, repeating the back-and-forth trip until the bikes meet, whereupon the bee is squashed.

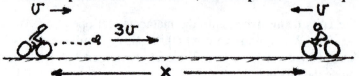

(a) What cumulative distance did the bee travel in its total back-and-forth trips?

Focus: $d = ?$

From $\bar{v} = \dfrac{d}{t}$, $d = \bar{v}t = 3vt$ for the bee.

We're told the bee's average speed is $3v$, but we're not told the time. The key to solving this problem is realizing that the time the bee spends flying is the *same* as the time it takes the bicycles to meet. Each bicycle goes a total distance $\frac{x}{2}$ in time t.

From $\dfrac{x}{2} = vt$, time $t = \dfrac{x}{2v}$. So the bee travels a total distance $d = 3v\left(\dfrac{x}{2v}\right) = \mathbf{1.5x}$.

(b) Calculate the cumulative distance of flight for the bee to the squish point, given that the bikes each travel at 10 km/h and the bee begins its 30-km/h back-and-forth trips when the bikes are 20 km apart.

Solution: As above, $d = 1.5x = 1.5(20 \text{ km}) = \mathbf{30\ km}$.

Check with all figures:

$$d = 3v\left(\frac{x}{2v}\right) = 3\left(10\,\tfrac{\text{km}}{\text{h}}\right)\left(\frac{20 \text{ km}}{2\left(10\,\tfrac{\text{km}}{\text{h}}\right)}\right) = \mathbf{30\ km.}$$

(Note that we can work directly with the units kilometers and hours without converting to meters and seconds.) The time of travel isn't given in the problem, but the equation for distance traveled (average speed multiplied by time) alerts us to *time* being the important factor. Trying to solve this problem without considering time is a stumper. Let the terms in the equations guide the steps to the solution.

Now have a go at the problems that follow!

Problems

3-1. The world's fastest train is presently the magnetically levitated Transrapid's Shanghai Maglev in China.
 (a) Write an equation for the average speed of the train if it travels distance x in time t.
 (b) Calculate the train's average speed in m/s if it travels 30.0 km in 8.0 minutes. Then convert to mph (mi/h).

3-2. A tennis ball is served and travels the length of the court L in time t.
 (a) Write an equation for the ball's average horizontal speed.
 (b) Calculate the average speed of a ball traveling 24.0 m across the court in 0.60 s.

3-3. A baseball pitcher throws a fastball across home plate. The ball crosses the plate, which has a front-to-back length x, in time t.
 (a) Write an equation for the speed at which the baseball crosses the plate.
 (b) Calculate the speed of a baseball that takes 0.010 s to cross home plate, 0.30 m from front to back.

3-4. A racecar races on a circular racetrack of radius r.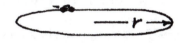
 (a) What is its average speed when it travels a complete lap in time t?
 (b) Calculate the average racing speed given that the radius of the track is 400 m and the time to make a lap is 40 s.

3-5. The new Taipei tower in Taiwan, of height h, has the world's fastest elevators.
 (a) How much time does it take an elevator rising from the ground floor to reach the top when the average elevator speed is $\bar{v}$?
 (b) Calculate the time of the upward ride to a height 508 m at an average speed of 15 m/s.
 (c) The elevators are said to zip upward at peak speeds of about 16 m/s, greater than the average speed. Is it reasonable that peak speeds can be 16 m/s when average speed is only 15 m/s? Defend your answer.

3-6. Phil runs the length of an American football field, 100 yards long.
 (a) How much time is required for him to run the full length at speed v (in m/s)?
 (b) Calculate the time it takes a person running at 6.0 m/s to run the length of an American football field.

3-7. Light is incredibly fast and travels at speed c. Consider light traveling along a ruler of length L.
 (a) How long does it take light to travel the length of the ruler?
 (b) Calculate the time it takes for light to travel the length of a meterstick. (The speed of light is 3.00×10^8 m/s.) Give your answer in nanoseconds. (1 nanosecond = 10^{-9} s.)

3-8. Lillian rides her bicycle along a straight road at an average velocity v.
 (a) How far does she travel in time t?
 (b) Calculate the distance covered by Lillian if her average speed is 7.5 m/s for a time of 5.0 minutes.

3-9. A gecko at rest sprints to a speed v in time t.
   (a) Assuming uniform acceleration, what is the average speed of the gecko?
   (b) What distance does the gecko cover in its sprint?
   (c) Calculate the distance the gecko covers when sprinting from rest to 2.0 m/s in a time of 1.5 s.

3-10. A skier starts down a slope from rest and reaches speed v in a time t.
   (a) Assuming a uniform pickup of speed, how far does the skier travel in this time?
   (b) Calculate how far the skier travels down the slope when starting from rest and reaching a speed 12 m/s in 8.0 s.

3-11. The cheetah is the fastest sprinter of all land animals. Suppose it starts from rest and accelerates uniformly to a speed v in a time t.
   (a) What distance will it cover?
   (b) Calculate the distance a cheetah covers if it starts from rest and accelerates uniformly to a speed of 100.0 km/h in a time of 8.0 seconds.

3-12. A moving van increases its speed from v_1 to v_2 in a time interval t.
   (a) What is the average acceleration of the van?
   (b) Calculate the average acceleration in m/s^2 that the van undergoes speeding up uniformly from 15 km/h to 40 km/h in 20 s.

3-13. A hybrid automobile traveling at speed v_1 uniformly increases to speed v_2 in a time interval t.
   (a) What is the average acceleration of the automobile?
   (b) Calculate the average acceleration in m/s^2 when the auto increases its speed from 5.0 km/h to 20.0 km/h in 10.0 s.
   (c) Calculate the distance the auto travels during this period of acceleration.

3-14. Lonnie applies the brakes to his car moving at speed v. The car slows at a constant rate and is brought to rest in time t.
   (a) What is the acceleration?
   (b) Calculate the acceleration if the initial speed of the car was 25 m/s and the time to stop was 20 s.
   (c) Calculate the distance traveled while the car was decelerating.
   (d) Calculate the distance the car travels between the instant Lonnie sees reason for braking and applying the brakes. Assume a 25-m/s speed and a reaction time of 1.5 s.

3-15. A jet plane lands on a runway with a speed v and after time t comes to a stop.
   (a) Assuming that its speed is reduced at a constant rate, find the acceleration of the plane.
   (b) Calculate the acceleration if the landing speed is 72 m/s and the stopping time is 12 s.
   (c) Calculate the distance the jet travels between the point of touchdown and the point of stopping.

3-16. A dart leaves the barrel of a blowgun at a speed v. The length of the blowgun barrel is L. Assume that the acceleration of the dart in the barrel is uniform.
 (a) For how long a time is the dart moving inside the barrel?
 (b) Calculate the time in the barrel if the dart's exit speed is 15.0 m/s and the length of the blowgun is 1.4 m.

3-17. A bullet leaves the barrel of a gun at a speed v. The length of the gun barrel is L. Assume that the acceleration of the bullet in the barrel is uniform.
 (a) What was the bullet's average speed inside the barrel?
 (b) Calculate the average speed if the bullet leaves the gun at 350 m/s. The length of the gun barrel is 0.40 m.
 (c) Calculate the time the bullet is in the barrel.

3-18. Stan brakes his car and slows at a uniform rate from v_0 to v in a time t.

 (a) How far does the car travel while slowing to the lower speed?
 (b) Calculate the distance the car travels while braking from 25 m/s to 11 m/s in a time of 8.0 s.

3-19. Motion picture frames show that a tossed ball moves distance x between frames. The rate at which the frames are taken is 24 per second.
 (a) How fast was the ball moving?
 (b) Calculate the speed of a ball that moves 0.40 m in each frame.

3-20. An electron placed in an electric field accelerates uniformly from rest to speed v while traveling a distance x.
 (a) What is the acceleration of the electron?
 (b) Calculate the acceleration in m/s^2 for an electron that starts from rest and reaches a speed of 1.8×10^7 m/s over a distance of 0.10 m.
 (c) Calculate the time required for the electron to attain this speed.

3-21. The speed of a toy rocket shooting straight upward increases from speed v to speed V at a uniform rate in a time t.
 (a) How far does the rocket travel during this time?
 (b) Calculate the distance (in meters) covered if the initial rocket speed is 110 m/s and increases uniformly to 250 m/s in a period of 3.5 s.

3-22. Rog tosses a ball straight upward at speed v. Ignore air drag.
 (a) How long does the ball take to reach its highest point?
 (b) Calculate the time in seconds that it takes for the ball to reach its zenith when thrown straight upward at 32 m/s.
 (c) Calculate the maximum height of the ball.

3-23. George drops a stone from atop a cliff of height h that overlooks the ocean.
 (a) Ignoring air drag, how long a time does the stone take to hit the water?
 (b) Calculate the number of seconds to hit the water if the stone is dropped from a cliff 25 m high.
 (c) Calculate the speed of the stone when it hits the water.

3-24. A motorist drives distance x from one city to another in time t, but makes the return trip in $0.75t$.
  (a) What is the average speed for the total trip?
  (b) Calculate the average speed for the round trip between cities 140 km apart if the initial outward trip takes 2.0 hours.

3-25. To reach your cabin, Tsing walks at an average speed v for 30 minutes, then jogs at $2v$ for another 30 minutes.
  (a) What is Tsing's average speed for her trip to the cabin?
  (b) Calculate the average speed for the whole trip for a walking speed of 1.0 m/s.
  (c) Calculate the distance between her starting point and the cabin.

3-26. Janet throws a ball straight upward. The ball soon returns to her hand.
  (a) How fast should she throw it so that it returns to her hand a time interval t later?
  (b) Calculate the throwing speed straight upward for a 4.0-second time of flight.
  (c) Calculate the maximum height reached by the ball.

3-27. The ceiling of a school gymnasium is distance y above the floor.
  (a) With what maximum upward speed can you throw a ball from an elevation 2.0 meters above the floor so that it just misses hitting the ceiling?
  (b) Calculate the maximum speed of the thrown ball if the floor-to-ceiling distance is 20.0 m.

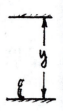

3-28. Dennis drives for 1 hour at average speed v. Then he drives for another hour at average speed $4v$.
  (a) Find the overall average speed.
  (b) Calculate his overall average speed if his average speed for the first hour is 25 km/h, and his average speed during the second hour is 100 km/h.

3-29. Suppose that you drive a distance x at an average speed v, then drive the same additional distance x at $1.5v$.
  (a) Find your average speed.
  (b) Calculate your average speed if you drive 1.0 kilometer at 28 km/h, then drive 1.0 kilometer back to your starting point at 42 km/h.

3-30. Judy and her dog Atti take their morning walk to the Vinoy Hotel, which is distance x away. Judy walks at a brisk speed v in a straight line while Atti runs back and forth at speed V between Judy and the hotel, until both reach the hotel.
  (a) What is the total distance that Atti runs?
  (b) Calculate the total back-and-forth distance that Atti runs if Atti's speed is 4.5 m/s, Judy's speed is 1.5 m/s, and the distance Judy walks is 150.0 m.

Show-That Problems for Linear Motion

The following problems supply numerical values, and the answer to each problem. What's important is for you to show how you arrive at the stated answer. Sometimes more information is supplied than required. (Aren't problems in everyday life also accompanied by extraneous information? ☺) Please continue with the practice of finding the solution with symbols before you plug in numerical values.

3-31. A ball rolls down a 3-m inclined plane in 1.5 s.
Show that the average speed is 2 m/s.

3-32. A ball with a temperature of 22°C is thrown vertically upward at 14.7 m/s.
Show that its maximum height will be 11 m.

3-33. A car traveling along a level road uniformly increases its speed from rest to 28.0 m/s in 8.0 s.
Show that it travels a distance of 112 m in this time.

3-34. A 40-gram egg falls from rest atop a 15-m tall tree.
Show that it takes 1.7 s for it to hit the ground below.

3-35. A ball rolling down an inclined plane starts from rest and reaches a speed of 12 m/s in 3 s.
Show that it has an acceleration of 4 m/s^2.

3-36. An F-14 Tomcat fighter jet goes from rest to a speed of 75 m/s in 2.5 s.
Show that the acceleration it undergoes is 30 m/s^2.

3-37. A motorboat accelerates from rest in a straight line at a constant 2.0 m/s^2 for a time of 8.0 s.
Show that it travels 64 m in this time.

3-38. Fred goes down a 5.0-m slide at a playground in 2.0 s. He starts from rest and accelerates uniformly.
Show that his acceleration while on the slide is 2.5 m/s^2.

3-39. Zephram is on roller blades on a hill. He starts from rest and skates for 5.5 s accelerating uniformly at 3.5 m/s^2.
Show that the distance traveled is 53 m.

3-40. A red-colored ball tossed upward reaches a maximum height of 3.0 m.
Show that its initial speed was 7.7 m/s.

3-41. Will tosses a golf ball straight upward.
Show that if the ball is tossed upward at 18 m/s it will be in the air for less than 4 s.

3-42. Carmelita tosses a baseball straight up into the air and it returns to her glove in 3.0 s.
Show that she has to give the ball a vertical speed of about 15 m/s if it is to return to her glove in that time.

3-43. When some volcanoes erupt, big rocks have been measured to shoot upward with speeds of about 1000 m/s.
Show that, neglecting air resistance, these rocks could reach heights exceeding 50 km.

3-44. A grizzly bear has a top running speed of about 13 m/s (about 30 mi/h). Two campsites are 65 m apart.
Show that a grizzly bear can run from one campsite to the other in 5 s.

3-45. A spiffy sports car accelerates from rest to 28 m/s (100 km/h) at an average acceleration of 7.0 m/s².
Show that this acceleration takes 4.0 s.

3-46. A car accelerates uniformly from rest to a speed of 25 m/s in a time of 5.0 s.
Show that the car covers a distance of 63 m.

3-47. Phil and Mala run a 100-m race and Mala gloriously wins in 12.8 s while Phil takes 13.6 s.
Show that Mala wins the race by a distance of 6 m.

3-48. When taking one of those fantastic shots moving toward the basket at a running speed of 3 m/s, Terrence rises about 1 m above the court floor.
Show that his hang time is 0.9 s.

3-49. At 60 mph it takes 60 s to travel a mile. At twice this speed, 120 mph, it takes 30 s.
Show that to cover a mile in 45 s requires a speed of 80 mph.

3-50. Norma drives to a destination at an average speed of 40 km/h and returns at an average speed of 60 km/h.
Show that her average speed is 48 km/h (and not 50 km/h!) for the round trip.

4   *Newton's Second Law*

Here's where dynamics begins, applying the concepts introduced in Chapter 3—*speed, velocity,* and *acceleration*. Acceleration is $\Delta v/\Delta t$, as defined previously. In this chapter we state in Newton's second law how acceleration is *produced*, $a = F_{net}/m$. We introduce the concept of mass. We build on the equilibrium rule of Chapter 2—when $\Sigma F = 0$ no acceleration occurs. We see that when acceleration does occur, then $\Sigma F = ma$. Since force and acceleration are vector quantities, we often treat the horizontal and vertical components separately: $\Sigma F_x = ma_x$ and $\Sigma F_y = ma_y$, where x and y denote the horizontal and vertical directions, respectively. The expressions F_{net} and ΣF are one and the same, ΣF being more suggestive of more than a single force acting to produce the acceleration.

The standard symbol for force is F. For more specific forces we use other symbols—f stands for friction, N for normal force, T for the force due to tension, and mg or W for weight, the force due to gravity. (In Chapter 9 we'll define weight as a support force. In a non-accelerating system, the weight and the support force have the same magnitudes.)

There are only a few *fundamental* units we will encounter in this course: the *meter* (length), the *kilogram* (mass), the *second* (time), the *kelvin* (temperature) and the *ampere* (current). All of the other units you'll see in this book are *derived* from these. For instance, 1 Newton is defined as the amount of force on a 1-kg mass that will produce an acceleration of 1 m/s², so $1 \text{ N} \equiv 1 \text{ kg·m/s}^2$.

Units of measurement are displayed in standard typeface, while symbols are shown in italics. For example, we can say a normal force $N = 5$ N, or a block of mass m travels a distance of 5 m. In the text and in the *Practicing Physics* book, $g = 10$ m/s², since multiples of 10 simplify mental math more than multiples of 9.8. In the problems in this book we use the more precise $g = 9.8$ m/s².

From Newton's second law $\left(a = \dfrac{F}{m} \right)$ and the force due to gravity as weight W,

$$g = \frac{W}{m} = \frac{9.8 \text{ N}}{1 \text{ kg}} = 9.8 \frac{\text{N}}{\text{kg}} = 9.8 \frac{\text{kg·}\frac{\text{m}}{\text{s}^2}}{\text{kg}} = 9.8 \frac{\text{m}}{\text{s}^2}.$$ We see that the units $\dfrac{\text{N}}{\text{kg}}$ and $\dfrac{\text{m}}{\text{s}^2}$ are equivalent and

we can use either for g. Throughout this book, g is the acceleration due to gravity. It can also signify *gravitational field strength*—gravity's pull per kilogram on an object at or near Earth's surface (Chapter 9). When used this way it is common to boldface g.

Problem-Solving Strategy

Simple Sample Problem 1
Find the acceleration of a crate of mass m on a horizontal surface when acted upon by a net force F.

Focus: $a = ?$

> The physics underlying this situation is Newton's second law. In equation form the solution is,

$$a = \frac{F_{net}}{m}.$$

This is a *one-step problem*. If we're given the values of F and m, the solution is a straightforward *One-step Calculation* as in the textbook.

Two-Steps Sample Problem 2

Find the acceleration of a crate of mass m on a factory floor when the horizontal net force F acting on it is equal to one-half the weight of the crate.

Step 1: *Focus*: $a = ?$

Step 2: The physics is still Newton's second law. In equation form:

$$a = \frac{F_{net}}{m} = \frac{\left(\frac{mg}{2}\right)}{m} = \frac{g}{2} = \mathbf{0.5g}.$$

Note that we've simply substituted $\frac{mg}{2}$ for F. Then after m cancellation we have our solution (shown in bold face).

Suppose we were told "The mass of the crate is 200.0 kg, its color is bright red, the floor surface temperature 22°C, and its initial speed 2.0 m/s. The horizontal net force acting on the crate is one-half the weight of the crate. Calculate the acceleration of the crate."

Note that the color of the crate, its initial speed, and the temperature of the floor are all irrelevant. They're not part of the solution. Note also that the mass cancels. So

Solution: $a = 0.5g = 0.5(9.8 \text{ m/s}^2) = \mathbf{4.9 \text{ m/s}^2}.$

Three-Steps Sample Problem 3

Find the acceleration of the same crate of mass m on a horizontal surface when a horizontal pull P acts against a force of friction f that is equal to one-fifth the weight of the crate.

Step 1: *Focus*: $a = ?$

Step 2: $a = \frac{F_{net}}{m}$. We begin with Newton's second law. Continuing, $a = \frac{P - f}{m}.$

Note that we substitute $(P - f)$ for the net force. Continuing,

Step 3: We substitute $\frac{mg}{5}$ for f. Since P, m, and g are known values, the problem is solved:

$$a = \frac{P - \frac{mg}{5}}{m}. \quad \text{(Or going an extra step we get } a = \frac{P}{m} - 0.20g.\text{)}$$

For a numerical solution, assume a mass of 200.0 kg and a horizontal pull of 600.0 N. Then the answer is:

$$a = \frac{600.0 \text{ N} - \frac{\left(200.0 \text{ kg} \times 9.8\frac{m}{s^2}\right)}{5}}{200.0 \text{ kg}} = 1.04 \frac{m}{s^2}.$$

Or equivalently, $a = \frac{600.0 \text{ N}}{200.0 \text{ kg}} - 0.20 \times 9.8 \frac{m}{s^2} = 1.04 \frac{m}{s^2}.$

Vertical-Acceleration Sample Problem 4

A box of mass m hangs by a string from the ceiling of an elevator.

(a) What is the tension T in the string when the elevator accelerates upward with acceleration a?

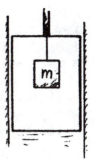

Focus: $T = ?$ Two forces act on the box—the tension force T and the gravitational force mg. If we call upward the positive direction, then

from $a = \dfrac{F_{net}}{m} = \dfrac{T - mg}{m} \Rightarrow T = mg + ma = m(g + a)$.

If the mass were just hanging and the elevator was not accelerating, T would equal mg. The quantity ma represents the additional force the string must provide to accelerate the box.

(b) Calculate the tension in the string if the mass of the box is 5.0 kg and the upward acceleration of the elevator is 1.4 m/s².

Plugging in, $T = m(g + a) = 5.0 \text{ kg}\left(9.8\frac{\text{m}}{\text{s}^2} + 1.4\frac{\text{m}}{\text{s}^2}\right) = \mathbf{56\ N}$.

(c) If the 5.0-kg box were instead resting on the floor of the same elevator moving upward with the same acceleration, what would be the normal force on the box?

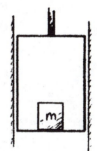

Answer: The normal force would be 56 N, the same as the tension that produces the same upward acceleration.

(d) Suppose the same box rests on the floor of the elevator while the elevator accelerates *downward* at 1.4 m/s². Calculate the magnitude of the normal force acting on the box.

Focus: $N = ?$

From $a = \dfrac{F_{net}}{m} = \dfrac{N - mg}{m} \Rightarrow N = mg + ma = m(g + a) = 5.0 \text{ kg}\left(9.8\frac{\text{m}}{\text{s}^2} + (-1.4\frac{\text{m}}{\text{s}^2})\right) = \mathbf{42\ N}$.

In this case the acceleration is negative because it is downward, opposite the direction we identified as positive. Notice that the normal force is less when the elevator accelerates downward. This makes sense, for in the extreme case of the elevator accelerating downward at g, free fall, the normal force would be zero.

(e) What would be the tension in part (a) and the normal force in part (c) if instead the elevator were moving at constant velocity?

Answer: At constant velocity, $a = 0$ and the supporting forces would simply be mg.

Two-Block Vertical and Horizontal Acceleration
Sample Problem 5

A cart of mass *m* on a horizontal friction-free air track is accelerated by a string attached to a weight also of mass *m* hanging vertically from a small pulley as shown.

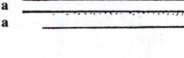

(a) Find the acceleration of the two-mass system.

Step 1: *Focus*: $a_{system} = ?$

Step 2: *Solution*: Newton's second law underlies the solution. The net force that accelerates the two-mass system is the force of gravity on the hanging weight, *mg*, so

$$a_{system} = \frac{F_{on\ system}}{m_{system}} = \frac{mg}{2m} = \frac{g}{2}.$$

(b) Show that the string tension provided by the dangling mass is $\frac{mg}{2}$ (and NOT *mg*!).

Step 1: *Focus*: Tension $T = ?$

Step 2: Newton's second law again. Consider the forces that act on the hanging weight: *mg* pulls down and *T* pulls up. In this case we'll consider acceleration to be positive when the cart moves to the right and the weight drops. From $\Sigma F = ma$ for the falling weight we get,

$$mg - T = \frac{mg}{2} \ \left(\text{since } a = \frac{g}{2} \right) \ \Rightarrow T = \frac{mg}{2}.$$

It's very important to realize that string tension is *less* than *mg* because otherwise the tension acting up would balance the gravitational force acting down and there would be no acceleration. If you held the sliding weight to prevent it from moving, then there would be no acceleration and *T* would be *mg*. Got it?

(c) Suppose instead of being equal, the mass on the track is 100.0 kg while the hanging mass is 1.00 kg. Calculate the acceleration when the 1.00-kg mass dangles over the pulley.

Step 1: *Focus*: $a_{system} = ?$

Step 2: Again, Newton's second law. Let's call the dangling mass *m* and the mass on the track *M*.

$$a_{system} = \frac{F_{on\ system}}{m_{system}} = \frac{mg}{(m + M)} = \left(\frac{m}{m + M} \right) g = \left(\frac{1.00\,\text{kg}}{1.00\,\text{kg} + 100.0\,\text{kg}} \right) 9.8\,\tfrac{m}{s^2} = \tfrac{1.00}{101.0} \left(9.8\,\tfrac{m}{s^2} \right) = 0.097\,\tfrac{m}{s^2}$$

(d) Interchange the 1.00-kg and 100.0-kg masses, so the 100.0-kg mass dangles over the pulley. What acceleration occurs now?

$$a_{system} = \left(\frac{m}{m + M} \right) g = \left(\frac{100.0\,\text{kg}}{1.00\,\text{kg} + 100.0\,\text{kg}} \right) 9.8\,\tfrac{m}{s^2} = \tfrac{100.0}{101.0} \left(9.8\,\tfrac{m}{s^2} \right) = 9.7\,\tfrac{m}{s^2}.$$

Notice how entirely different the accelerations are. This leads to the next question.

(e) For a variety of masses for this experiment, what range of accelerations is possible? (Hint: Let one mass be nearly zero (like a feather) and the other mass huge; then switch.)

Answer: The range of possible accelerations is from **0 to g**. From $a_{\text{system}} = \left(\dfrac{m}{m+M}\right)g$, if

m (the hanging mass) is very small and M (the pulled mass) is very large, the ratio $\left(\dfrac{m}{m+M}\right)$ is nearly zero, and acceleration is nearly zero. If m is very large

and M is small, the ratio $\left(\dfrac{m}{m+M}\right)$ is nearly one, so acceleration is nearly g

(which makes sense, since a heavy weight pulling a feather across the track is essentially in free fall. These scenarios are featured on pages 17 and 18 of the Practice Book).

Problems

4-1. Suzie exerts a horizontal force F on a box of mass m that sits on a frictionless horizontal surface.
(a) Write an equation for the acceleration of the box.
(b) Calculate the acceleration when her push is 15 N and the mass of the box is 10 kg.

4-2. The engines of a Boeing 747 jet of mass m produce an acceleration a along the runway during takeoff.
(a) Write an equation for the net force that acts on the jet.
(b) Calculate the force when the loaded jet has a mass 3.2×10^5 kg and acceleration of 3.5 m/s² down the runway.

4-3. When using a spring balance in lab you find that a force F accelerates a block across a nearly frictionless horizontal surface at acceleration a.
(a) Find the mass of the block.
(b) Calculate the mass of the block if a 5.0-N force accelerates it 2.5 m/s².

4-4. Helen exerts a horizontal force F on a puck of mass m on a horizontal surface where the force of friction is f.
(a) What is the resulting acceleration?
(b) Calculate the acceleration when her push is 5.0 N, the mass of the puck is 0.50 kg, and the friction is 1.0 N.

4-5. A motorcycle of mass m speeds up from speed v_0 to speed v_f in a time t.
(a) What is the average force F acting on the motorcycle?
(b) Calculate the force that will accelerate a 210-kg motorcycle from 12.5 m/s to 20.8 m/s in 6.9 seconds.

4-6. A certain net force applied to a vehicle of mass m causes it to accelerate at a.
(a) How much acceleration will the same force produce on a vehicle of mass 1.5 m?
(b) Calculate the acceleration of the heavier vehicle if the lighter one has a mass of 4000 kg and the applied force on each is 6000 N.

4-7. In serving, a tennis player accelerates a tennis ball of mass m horizontally from rest to a speed v. The acceleration is uniform when applied by the racquet over time t.

(a) How much force is exerted on the ball by the racquet?

(b) Calculate the average force when the mass of the ball is 56 grams and its speed from the racquet is 32 m/s for an impact time of 0.090 s.

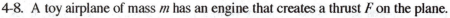

Note that since our force unit is kg·m/s², the mass you plug in needs to be in kilograms.

4-8. A toy airplane of mass m has an engine that creates a thrust F on the plane.

(a) What is the acceleration of the plane moving horizontally when air drag is R?

(b) Calculate the acceleration if the mass of the plane is 1.1 kg, the engine produces a forward force of 18 N, and air drag is 8 N.

4-9. When a horizontal force F is applied to a box having a mass m, the box slides on a level floor, opposed by a force of kinetic friction f.

(a) What magnitude of acceleration occurs for the box?

(b) Calculate the acceleration if the horizontal force is 412 N, friction is 122 N, and the mass of the box is 75 kg.

4-10. Car A has mass m and experiences a net force F. Car B has mass $m/2$ and experiences a net force $F/2$.

(a) Which car has the greater acceleration?

(b) Calculate the accelerations when $m = 2000$ kg and $F = 4000$ N.

4-11. Jean produces a net force F on a cart of mass m that gives it an acceleration a.

(a) How much acceleration will the same force produce on a cart of 0.60 m?

(b) If the mass m of the original cart is 4.0 kg and its acceleration is 2.0 m/s², how much acceleration will the same force produce on the lighter cart?

4-12. Two boxes with masses m and M rest on a horizontal surface. Each box is mounted on tiny rollers and moves with negligible friction. A force applied to m gives it an acceleration a.

(a) What acceleration will M undergo if pushed by the same amount of force?

(b) The two boxes have masses 2 kg and 5 kg. A force on the lighter one gives it an acceleration of 2 m/s². Calculate the acceleration of the 5-kg box when it is pushed by the same amount of force.

4-13. A car of mass m has a maximum acceleration a. When it pulls a trailer of mass $2m$, its acceleration is less.

(a) Find the maximum acceleration of the car and trailer together.

(b) A friend says the answer is $a/2$. Why should you suggest that your friend go back and try again?

4-14. A stone of weight W rests on a horizontal friction-free surface.
   (a) What acceleration occurs when a horizontal force F acts on the stone?
   (b) Calculate the horizontal acceleration on the friction-free surface for a 75-N horizontal force on a stone that weighs 38 N.

4-15. George pushes a stone of weight W upward with a force F that is greater than the stone's weight.
   (a) What vertical acceleration occurs?
   (b) Calculate the acceleration when the weight of the stone is 38 N and the upward push is 75 N.

4-16. A bucket of water of mass m is pulled horizontally on a friction-free surface by a force equal to 3 times the weight of the bucket.
   (a) What is its acceleration?
   (b) What would be the acceleration of the bucket if it were pulled with a force equal to 4 times the bucket's weight?

4-17. A bucket of water of mass m is pulled straight upward by a force equal to 3 times its weight.
   (a) What is its acceleration?
   (b) A friend says the answer is $3g$. What error did your friend likely make?

4-18. A paint bucket of mass m initially at rest is pulled upward with a steady force P that is greater than the weight of the bucket.
   (a) What is the net force acting on the bucket?
   (b) What is the acceleration of the bucket?
   (c) Calculate the acceleration if the mass of the paint bucket is 8.5 kg and it is pulled upward with a force of 95 N.

4-19. When a parachutist of mass m falls, the upward drag force on the chute is R.
   (a) What is the acceleration of the parachutist?
   (b) R is greater than mg when the parachute first opens. What then is the direction of the acceleration? What is the direction of the velocity?
   (c) Calculate the acceleration of the falling parachutist when the drag force is 1000 N and the mass of the parachutist is 80 kg. Neglect the mass of the parachute.
   (d) How large a drag force will produce terminal velocity?

4-20. George drops a stone of mass m from atop a high cliff of height h that overlooks the ocean. Just before the stone hits the water, air drag has built up to equal exactly one-fifth the weight of the stone.
   (a) What is its acceleration at this point (expressed as a fraction of g)?
   (b) Calculate the acceleration of the stone in m/s^2 just before it hits the water.

4-21. A small puck of mass m rests on a horizontal air table with negligible friction.
   (a) How much horizontal force must be applied to the puck to make it accelerate as much as it would if it were in free fall?
   (b) If the mass of the puck is 1.1 kg, calculate the horizontal force to give it an acceleration of g.

4-22. A subcompact car of mass m and a brawny pickup truck of mass $3m$ are given equal accelerations.
(a) How much greater is the force that acts on the more massive vehicle?
(b) How would the accelerations compare if the same force acted on each vehicle?

4-23. A braking force F slows a truck of mass m from speed v_o to speed v in a time t.
(a) How much braking force must be applied?
(b) Calculate the braking force if the mass of the truck is 3.0×10^3 kg and it is slowed from 32 m/s to 12 m/s in 6.0 seconds.

4-24. A loaded pick-up truck of mass m moves with speed v. Its brakes are applied, and it comes to a stop in t seconds.
(a) What average braking force slows the truck?
(b) What is the stopping distance x?
(c) Calculate the braking force if the truck's mass is 2.0×10^3 kg, its initial speed is 18 m/s, and the stopping time is 8.0 s.
(d) Calculate the stopping distance.

4-25. During a collision, a driver of mass m in a car moving at speed v is brought to rest by an inflated air bag in time t.
(a) What is the average force exerted by the air bag on the driver?
(b) Calculate the average force if the mass of the driver is 55 kg, the initial speed of the car 28 m/s, and the contact time with the air bag is 0.20 s.

4-26. An average force is applied by a shoulder-strap seatbelt to a passenger of mass m in an emergency stop when the car's initial speed is v and it stops in t seconds.
(a) How great is this average force?
(b) Calculate the average force on a 55-kg passenger when the car slows from 85 km/h to a complete stop in 3.2 s.

> Recall that the unit of force, N, is kg·m/s².
> Check your units before you plug in numbers

4-27. A jet plane of mass m catapults from rest to a speed v in time t from the deck of an aircraft carrier.
(a) What average net force acts on the jet?
(b) Calculate the average force that propels a jet plane of mass 26,500 kg from rest to a speed of 72 m/s in a time 1.9 s.

4-28. A certain force accelerates an electron of mass m from rest to a speed v in time t.
(a) How great is this force?
(b) Calculate the force that accelerates an electron of mass 9.1×10^{-31} kg from rest to a speed of 3.0×10^6 m/s in a time of 1.5×10^{-7} s.

4-29. A golf ball of mass m leaves the tee at a speed v, which it gains in a brief time t.
(a) What is the average force of impact of the club on the ball?
(b) Calculate the average force of the club on the ball of mass 0.045 kg required to give it a speed of 72.0 m/s in an impact time 0.0050 s.

4-30. Firefighter Fred of mass m slides down a vertical pole with an acceleration $0.25g$.
(a) What is the force of friction that acts on Fred?
(b) What friction force would be needed to slide down at constant speed?

4-31. Suzy Skydiver who has mass m steps from the basket of a high-flying balloon of mass M and does a sky dive.
(a) What is the net force on Suzy at the moment she has stepped from the basket?
(b) What is the net force on her when air drag builds up to equal half her weight?
(c) What is the net force on her when she reaches terminal speed v?
(d) What is the net force on her after opening her parachute and reaching a terminal speed $0.1v$?

4-32. A landing craft of mass m prepares itself for a Moon landing. When at a vertical distance y above the Moon's surface its speed downward is v. A retrorocket is fired to give the craft an upward thrust to slow its speed to zero as it meets the surface.
(a) How much is this thrust? Use g_{Moon} for the acceleration due to gravity near the Moon's surface.
(b) Calculate the needed thrust to decelerate the craft from a downward velocity of 15 m/s when 160 m above the lunar surface to rest at the lunar surface. The mass of the craft is 12,000 kg, and the acceleration due to gravity on the Moon is $g/6$.

4-33. An astronaut of mass m floating in space outside her spaceship receives a thrust F from a nitrogen spurt gun. The duration of the spurt is t.
(a) How fast will she be moving relative to the spaceship?
(b) Calculate her speed relative to the spaceship if her mass is 88 kg, the spurting force is 32 N, and the duration of the spurt is 2.1 s.

4-34. A cart of mass m is accelerated from rest by a net force F acting for a time t.
(a) What will be the speed of the cart at the end of that time?
(b) Calculate the final speed of the cart of mass 5.0 kg when a 2.0 N force acts for 10 s.

4-35. An astronaut of mass m sits in a rocket that blasts off from rest attaining a speed v in time t.
(a) Assuming constant acceleration, what force does the spacecraft exert on the astronaut during takeoff? (This is the normal force, the astronaut's apparent weight.)
(b) The mass of the astronaut is 75 kg. What is the astronaut's weight before takeoff?
(c) Calculate the 75-kg astronaut's apparent weight during takeoff while the rocket gains a vertical speed of 42 m/s in 14 s.

4-36. Tammy applies a constant horizontal force F to a block of ice on a smooth horizontal floor. Friction is negligible. The block starts from rest and reaches a speed v in time t.
(a) What is the mass of the block of ice?
(b) If Tammy stops pushing at the end of 3.0 s when the speed of the block is 2.4 m/s, how far will the block slide in the next 3.0 s?

4-37. An iceboat of mass M with occupant of mass m begins at a rest position. A wind from directly behind exerts a force F on the sail.
(a) If friction is negligible, what is the acceleration of the iceboat?
(b) As the boat gains speed, the force of wind impact decreases. How does this affect acceleration?
(c) When the speed of the iceboat approaches wind speed, what will be the resulting acceleration of the iceboat?

4-38. A truck of mass M tows a car of mass m along a level road by a chain. The forward force on the truck's tires from the road is F.

(a) What is the acceleration of the car?
(b) What is the tension in the connecting chain?
(c) Calculate answers to the above questions if the mass of the truck is 3000 kg, the mass of the car is 1200 kg, and the driving force by the tires on the truck is 4800 N.

Show-That Problems for Newton's Second Law
(Please continue to find your solution with symbols before you plug in numerical values. And most of all, show how you arrive at your answer.)

4-39. A 60.0-kg astronaut finds his weight to be 96.0 N on the Moon.
Show that the acceleration due to gravity on the Moon is 1.6 m/s^2.

4-40. A net force of 10.0 N is exerted by Irene on a 2.0-kg cart for 3.0 seconds.
Show that the cart will have an acceleration of 5.0 m/s^2.

4-41. Toby Toobad, who has a mass of 100 kg is skateboarding at 9.0 m/s when he smacks into a brick wall and comes to a dead stop in 0.2 s.
Show that his deceleration is 45 m/s^2 (ouch!).

4-42. A 5.0-kg cart moving with a velocity of 4.0 m/s is brought to a stop in 2.0 s.
Show that the stopping force is 10.0 N.

4-43. A 1.0-kg ball traveling at a velocity of 45 m/s bounces from a wall with a velocity of -33 m/s, after an impact time of 0.20 s.
Show that the force of impact on the ball is 390 N.

4-44. A certain force exerted on an 8.0-kg block gives it an acceleration of 5.0 m/s^2.
Show that the same force exerted on a 2.0-kg block will give it an acceleration of 20 m/s^2.

4-45. A net force of 10.0 N on a block causes it to accelerate 2.0 m/s².
Show that the mass of the block is 5.0 kg.

4-46. Dan's truck has a mass of 2000 kg. When traveling at 22.0 m/s, it brakes to a stop in 4.0 s.
Show that the braking force acting on the truck is 11,000 N.

4-47. A 5.0-kg block accelerates at 0.80 m/s² along a horizontal surface when a horizontal 12.0-N force acts on it.
Show that the force of friction on the block is 8.0 N.

4-48. A falling 50-kg parachutist experiences an upward acceleration of 6.2 m/s² when she opens her parachute.
Show that the drag force is 800 N when this occurs.

Static Friction

Consider a wooden block at rest on a lab table. A string fastened to the side of the block passes over the pulley to a suspended cup. You (patiently!) add water to the cup a little bit at a time, and finally one drop at a time. Friction between the block and table surfaces holds the block in place. This is *static friction*—which occurs because of some adhesion as well as irregularities in the surfaces of the block and table. As more water is added to the cup, string tension increases. While the block is at rest, $\Sigma F = 0$, so increases in string tension are met with corresponding increases in the opposing static friction. A force diagram shows all the forces that act on the block.

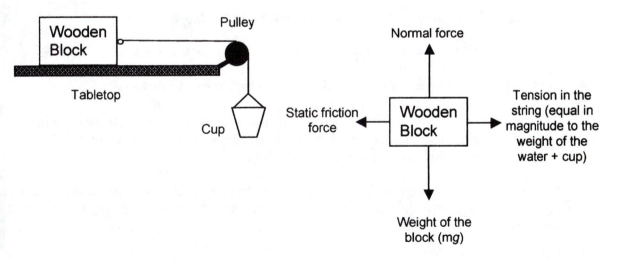

At a critical tension a threshold is overcome and sliding starts. That threshold of motion is characterized by a *coefficient of static friction* μ_s, defined as

$$\mu_s = \frac{f_{\text{static, max}}}{N}.$$

In other words, the coefficient of static friction for a given pair of surfaces equals the *maximum* value of static friction between the surfaces divided by the normal force N that squeezes them together. Note that μ_s is a dimensionless ratio, one force divided by another force.

Hence the force of static friction can be expressed

$$f_{\text{static}} \leq \mu_s N.$$

The "≤" symbol means that the static friction between the block and the table can range from zero (when nothing is pulling horizontally on the block) to some maximum value $(\mu_s N)$, beyond which the block starts to slide.

In the experiment above, if we were to add a heavy weight on top of the wooden block, the values of $f_{\text{static max}}$ and N would both increase proportionately so that the value of μ_s would not be affected. The coefficient of static friction, μ_s, is a constant for any two given materials in contact regardless of their individual size or weight. For example, the coefficient of static friction for two surfaces of wood is typically 0.4. For rubber in contact with wood it is about 1.0. A higher μ_s means that a larger force is required to get the surfaces to slide against each other.

Sample Problem 1

A 1.8-kg block of wood rests on a table. A 5-N force pulls horizontally on the block, but is not large enough to move the block.
(a) Draw a force diagram for the block.

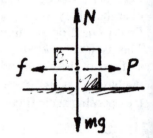

    Comments: Since the block is at rest, $\Sigma F = 0$. Upward and downward forces cancel exactly so we draw the N and the mg vectors with equal-lengths. In the horizontal direction, pull P must have the same magnitude as f_s.

(b) What is the magnitude of the static friction force on the block?

    Solution: The force of static friction must be equal in magnitude to the pulling force, **5 N**.

(c) What is the coefficient of static friction for the block and table surfaces?

    Answer: We can't determine μ_s from the data given. μ_s tells us the *maximum* value that the static friction force can have, but not the actual value of the friction force unless the block is at the threshold of sliding.

Suppose, however, that a pull of 7.0 N is just enough to break the block free. Static friction at this point is 7.0 N, and since the table is horizontal, the normal force equals the weight of the block, $mg = 17.6$ N. Then

$$\mu_s = \frac{f_{\text{static}}}{N} = \frac{7.0 \text{ N}}{17.6 \text{ N}} = \mathbf{0.40}.$$

We see that in this case that the maximum friction force before breakaway is 0.40 times the weight of the block. That's because the weight of the block and the normal have the same magnitude. On an inclined plane the normal would be less than the weight of the block and the block would have a greater tendency to slide. But that's another story—to be treated in Chapter 5.

Kinetic Friction

Returning to our block on the table. We replace the cup and pulley arrangement with a spring scale. Now we give the block a little tap to get it moving and record how much horizontal force we must exert to keep the block moving across the table at a constant speed.

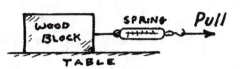

Experimentally, we find the horizontal force needed to maintain constant velocity is less than the initial force needed to start the block moving. Further experimentation will show that the pull needed to keep the block moving is approximately the same no matter how slowly or quickly the block moves (unless the speed is very high). The pull is also the same no matter which side of the block is in contact with the table (when on a narrow side, the effect of a smaller contact area is offset by a correspondingly greater pressure on that area). To a good approximation, friction doesn't depend on surface area. Since the block is moving at a constant speed we know that the net force on it is zero, so the friction force must be equal in magnitude to the pull.

We define the coefficient of kinetic (i.e., moving) friction by

$$\mu_k = \frac{f_{\text{kinetic}}}{N}$$

This coefficient, like the coefficient of static friction, depends only upon what the two surfaces are made of, not on the weight of the object or how big the area of contact is.

That means the force of kinetic friction can be written

$$f_{\text{kinetic}} = \mu_k N$$

This looks a lot like the static friction equation except that now there's an equal sign. The kinetic friction force has a fixed value for a given N, regardless of the speed of the motion. For wood on wood, μ_k has a typical value of about 0.3, less than the static coefficient of friction.

Sample Problem 2

Myrna must exert a horizontal 5.2-N force to pull a 2.5-kg box of chocolates across the dining room table at a constant speed.
(a) Make a force diagram of the box of chocolates.

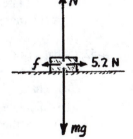

(b) What is the coefficient of kinetic friction between the box of chocolates and the table?

Focus: $\mu_k = ?$

Since the box is moving in a straight line at constant speed
$$\Sigma F_y = 0, \quad N - mg = 0 \Rightarrow N = mg, \text{ and}$$
$$\Sigma F_x = 0, \quad P - f_k = 0 \Rightarrow f_k = P.$$

From the equation for kinetic friction, with $f_k = P$ and $N = mg$,
$$\mu_k = \frac{f_k}{N} = \frac{P}{mg} = \frac{5.2 \text{ N}}{(2.5 \text{ kg})\left(9.8 \frac{\text{m}}{\text{s}^2}\right)} = \textbf{0.21.}$$

(c) What will be the acceleration of the box if Myrna exerts a more enthusiastic 7.3-N pull?

Focus: $a = ?$
The acceleration will be due to a net force in the horizontal direction.

Solution: $a_x = \dfrac{F_{net}}{m} = \dfrac{P - f_k}{m} = \dfrac{P - \mu_k mg}{m} = \dfrac{P}{m} - \mu_k g.$

(The first term in the solution is the acceleration the box would have with no friction. The second term is the reduction in the acceleration caused by the kinetic friction force.)

With our numbers: $a_x = \dfrac{7.3\left(\text{kg}\cdot\frac{\text{m}}{\text{s}^2}\right)}{2.5\ \text{kg}} - (0.21)\left(9.8\frac{\text{m}}{\text{s}^2}\right) = 0.86\frac{\text{m}}{\text{s}^2}.$

Coefficients of Friction Problems

4-49. A friction force f acts upon a horizontally moving box of weight W.
   (a) What is the coefficient of kinetic friction between the box and the surface?
   (b) Calculate the coefficient of kinetic friction between a moving box and the surface if the friction force is 22 N and the weight of the box is 56 N.

4-50. Bob finds that a constant horizontal pull P is needed to keep a crate of mass m moving across the floor at a constant velocity v.
   (a) What is the coefficient of kinetic friction μ_k between the crate and the floor?
   (b) Calculate μ_k if the pull is 87 N and the mass of the crate is 32 kg.

4-51. Emily exerts a horizontal force F to push a crate of oranges of mass m across a horizontal floor. The moving crate gains speed at a uniform rate a.
   (a) Find the coefficient of kinetic friction between the crate and the floor.
   (b) Calculate μ_k when the mass of the crate is 22 kg, the pushing force is 115 N and the crate increases speed at a rate of 0.80 m/s each second.

4-52. Megan finds that a constant horizontal pull P is needed to keep a box of mass m moving across the floor at a constant forward acceleration a.
   (a) What is the coefficient of kinetic friction between the box and the floor?
   (b) Calculate μ_k for the case of a horizontal pull of 143 N accelerating a 42-kg box across the floor at 1.3 m/s^2.

4-53. Alex finds that a constant horizontal pull P is needed to keep a box of mass m moving horizontally across the floor at a constant velocity v. The coefficient of kinetic friction between the box and the floor is μ_k.
   (a) What is the magnitude of P?
   (b) Calculate the horizontal pull to keep a 25-kg box moving horizontally across the floor at constant velocity when the coefficient of kinetic friction between the box and the floor is 0.30.

4-54. Ludmila exerts a constant horizontal pull P to keep a box of mass m moving across a level floor at a constant acceleration a. The coefficient of friction between the box and the floor is μ_k.

(a) What is the magnitude of P?

(b) Calculate the horizontal pull needed to make a 21-kg box slide across the floor with an acceleration of 0.50 m/s² if the coefficient of kinetic friction between the box and the floor is 0.35.

4-55. Stan the stuntman of mass m is dragged along a road at constant speed v by a cable attached to a moving truck of mass M. The cable is parallel to the level road and the coefficient of kinetic friction between Stan and road is μ_k.

(a) Make a force diagram of this scenario.

(b) Find the tension in the cable.

(c) If the truck has a mass of 3200 kg and it is dragging 82-kg Stan at speed 5.0 m/s, calculate the tension in the cable when μ_k between Stan and the road is 0.60.

(d) How would the tension differ if Stan were dragged at a constant speed of 6.0 m/s?

4-56. In order to hold a book of mass m stationary against a wall, Leslie finds she must push horizontally with a force of at least P.

(a) First make a force diagram, then find the coefficient of static friction.

(b) Calculate the coefficient of static friction for the wall and book when at least a 35-N horizontal push is needed to hold a 1.2-kg physics book steadily at rest against a wall.

4-57. The driver of a car of mass m moving at speed v slams on the brakes, which lock. The car then skids to a stop in a given distance x.

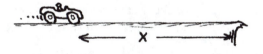

(a) What is the coefficient of kinetic friction between the tires and the road?

(b) Calculate μ_k for a 1500-kg car moving at 62 km/h that skids a distance of 28 m to rest when its brakes are locked.

4-58. • A crate of mass m rests on the flatbed of a truck of mass M. Coefficient of static friction between the crate and the surface of the flatbed is μ_k.

(a) What is the maximum forward acceleration the truck can have on a level road before the crate starts to slide backward relative to the bed of the truck?

(b) Once the truck and crate move at the same speed v, what is the maximum deceleration the truck can have before the crate slides forward relative to the bed of the truck?

(c) If the truck stops suddenly by hitting a wall, what is the maximum speed of the crate as it bashes against the cabin of the truck? (Why is Newton's first law sufficient to answer this, with no calculations needed?)

Show-That Problems for Coefficients of Friction

4-59. A 245-kg safe moves along a level floor. The coefficient of kinetic friction between the safe and the floor is 0.62.
Show that a horizontal force of about 1500 N will be required to pull it along at a constant speed.

4-60. While experimenting at lunchtime Marjorie finds that a 1.4 N horizontal force is needed to start a 340-gram sandwich sliding across the cafeteria table.
Show that the coefficient of static friction between the sandwich and the table is 0.42.

4-61. In an experimental mood Dave finds that a root beer mug slid across a bar top with an initial speed of 2.9 m/s slows down but is still moving when it reaches its intended recipient 4.0 m down the bar 1.6 seconds later.
Show that the coefficient of friction between the (wet) glass and the bar top is 0.051.

4-62. Steve slides a hockey puck at 8.5 m/s across the ice, which comes to rest 46 meters from its starting point.
Show that the coefficient of kinetic friction between the puck and the ice is 0.080.

4-63. A 130-gram calendar can be pushed horizontally against a wall with just enough force so that it doesn't slide down the wall. The coefficient of static friction between the calendar and the wall is 0.51.
Show that you'll have to exert at least a 2.5-N horizontal push to prevent it from sliding.

4-64. Atti, a 2.5-kg poodle, slides across a frozen lake at a speed of 6.3 m/s. The coefficient of kinetic friction between Atti and the ice is 0.088.
Show that Atti will slide for 7.3 seconds before coming to rest.

4-65. Consider the two-block acceleration problem (Sample Problem 5) on page 16. With no friction, both horizontal and vertical acceleration of the blocks are 0.5g. Suppose, however, that friction exists between the sliding cart and the track, with $\mu_k = 0.2$.
Show that the acceleration of the two-mass system will be 0.4 g.

Elementary Trigonometry

A Short Introduction To Trigonometry

Trigonometry blends of a bit of geometry with a lot of common sense. It lets you solve problems that would otherwise be undoable. The part of trigonometry that will be useful in this book deals with the relationships between the lengths of the sides of a right triangle.

Consider the following two similar right triangles:

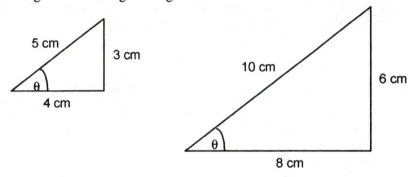

The symbol θ is the Greek letter "theta." We use θ to represent the size of the angle. Because the triangles are similar, θ is the same in both triangles.

It is an interesting fact (related to the similarity of the triangles) that the ratio of the length of the triangle's side opposite θ to the length of the hypotenuse gives the same result for both triangles:

$$\frac{side\,opposite\,\theta}{hypotenuse} = \frac{3\,cm}{5\,cm} = \frac{6\,cm}{10\,cm} = 0.6$$

EVERY right triangle that has an angle θ of this size has the same value for this ratio. We give this ratio, $\dfrac{side\,opposite\,\theta}{hypotenuse}$, a special name. We call it the "sine of theta", usually written sin θ.

Two other useful ratios are:

$$\text{Cosine of } \theta = \frac{side\,adjacent\,to\,\theta}{hypotenuse} = \frac{4\,cm}{5\,cm} = \frac{8\,cm}{10\,cm} = 0.8 \text{ in this example,}$$

and

$$\text{Tangent of } \theta = \frac{side\,opposite\,to\,\theta}{side\,adjacent\,to\,\theta} = \frac{3\,cm}{4\,cm} = \frac{6\,cm}{8\,cm} = 0.75 \text{ in this example.}$$

So we can think of the sine, cosine, and tangent as properties of the angle, since they are the same for every right triangle that contains a particular angle.

If we choose a different right triangle we get a different set of ratios:

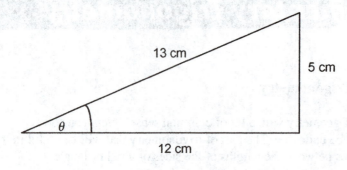

$$\sin\theta = \frac{5\ cm}{13\ cm}$$

$$\cos\theta = \frac{12\ cm}{13\ cm}$$

$$\tan\theta = \frac{5\ cm}{12\ cm}$$

It is useful to imagine that people have drawn every right triangle possible, and measured all of these ratios for every possible angle, and stored all of that information in your calculator (that's not how it's actually done, but it is useful to imagine as such).

Sample Problem 1

You lean your 7.0-foot ladder up against the wall such that it makes a 65° angle with the floor.

(a) How high up along the wall is the top of the ladder?

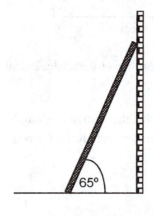
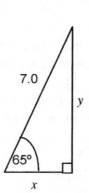

Focus: $y = ?$ We can model the ladder-against-the-wall as a right triangle.

$$\text{Solution: } \sin 65^\circ = \frac{y}{7.0\ \text{ft}}$$

$$\Rightarrow y = (7.0\ \text{ft})\sin 65^\circ = 7.0\,\text{ft} \times 0.906 = \textbf{6.3 ft}.$$

(b) How far from the wall is the base of the ladder?

$$\text{Solution: } \cos 65^\circ = \frac{x}{7.0\,\text{ft}}$$

$$\Rightarrow x = (7.0\ \text{ft})\cos 65^\circ = 7.0\ \text{ft} \times 0.423 = \textbf{3.0 ft}.$$

Sample Problem 2

**You are lying on the ground looking at a tree 45 meters away. You have to look up 32°
above the ground level to see the top of the tree.**
(a) How tall is the tree?

Focus: $y = ?$ (the side opposite the angle)
Model the situation as a right triangle.
We know the angle and the side adjacent to the angle.

From Tangent $\theta = \dfrac{\text{side opposite to } \theta}{\text{side adjacent to } \theta}$, rearranging

gives side opposite to $\theta =$
tangent $\theta \times$ side adjacent to θ.

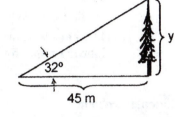

Solution: $y = (\tan 32°)(45\text{m}) = (0.625)(45\text{m}) = \textbf{28 m}$.

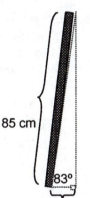

Sunlight from directly overhead at noon

**(a) You put a post into the ground a little crooked, making an angle
of 83° with the ground. The protruding bit of the post is 85 cm
long. When the Sun is directly overhead at noon, how long is the
shadow of the post?**

Focus: $x = ?$ (length of the shadow of the post)
Again, draw a right triangle. We know the hypotenuse and
the angle.

From cosine $\theta = \dfrac{\text{side adjacent to } \theta}{\text{hypotenuse}}$. Rearranging gives

side adjacent to $\theta = $ cosine $\theta \times$ hypotenuse.

Solution: $x = (\cos 83°)(85\text{cm}) = (0.122)(85\text{cm}) = \textbf{10 cm}$.

85 cm

$x =$ Length of
the shadow of
the post

Displacement

Sample Problem 3

**Suppose you are out in the desert in search of a
town. The closest town is 5.0 km north of you, and
the town that you want is 8.0 km west of that.
Instead of going first to the town north of you and
then to the second town, you decide to take a
shortcut directly to the second town.**
(a) In what direction should you proceed?

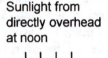

Solution: From the diagram our θ is
"the angle whose tangent is 1.6"
$\left(\tan \theta = \frac{8.0 \text{ km}}{5.0 \text{ km}} = 1.6 \right)$. This is called the
inverse tangent (or sometimes the *arc tangent*) of 1.6. It is written as
$\tan^{-1} 1.6$ *or* arctan1.6. Your calculator should give you an answer of **58°**.

(b) How much distance do you "save" by taking the shortcut?

Solution: The distance you save is the difference between how far you walk without the shortcut and how far you walk with the shortcut. Without the shortcut you walk (5.0 km + 8.0 km) = 13.0 km. To find the distance taking the shortcut use the Pythagorean theorem, $a^2 + b^2 = c^2 \Rightarrow c = \sqrt{a^2 + b^2}$.

The shortcut distance is $\sqrt{(5.0 \text{ km})^2 + (8.0 \text{ km})^2} = 9.4$ km, which saves you (13.0 km – 9.4 km) = **3.6 km** of hiking through the desert.

Knowing some trig could save you an hour's walk in the hot sun!

Sample Problem 4

Suppose you come to a wide river that moves at a steady speed of 3 mi/h. A man will rent you a boat that can go 7 mi/h. The love of your life is waiting for you, directly across the river. In which direction do you steer the boat so that you can arrive there as directly as possible?

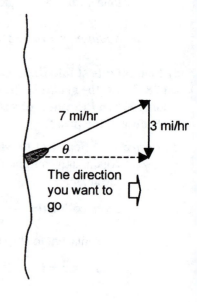

7 mi/hr

3 mi/hr

θ

The direction you want to go

Solution: If you were to run the boat for 1 hour, you'd travel 7 miles relative to the river, but the river would carry you 3 miles downstream. You want to end up going straight across.

From the diagram the angle you want has a sine equal to 3/7. That is,

$$\sin\theta = \frac{3}{7} \quad \Rightarrow \quad \theta = \sin^{-1}\left(\frac{3}{7}\right) = \mathbf{25°}.$$

Now have a go at the problems that follow!

Problem 1: Given a side and an angle in diagrams (a) and (b), find the other 2 sides.

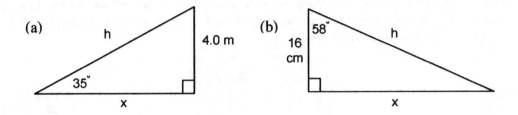

Problem 2: Given 2 sides in diagram (c) and one side in (d), find the missing angles and sides.

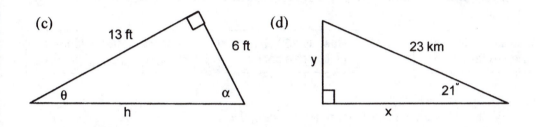

Solutions: 1.(a) $x = ?$ $h = ?$

$$\tan 35° = \frac{4.0\text{ m}}{x} \quad \Rightarrow x = \frac{4.0\text{ m}}{\tan 35°} = \textbf{5.7 m.}$$

$$\sin 35° = \frac{4.0\text{ m}}{h} \quad \Rightarrow h = \frac{4.0\text{ m}}{\sin 35°} = \textbf{7.0 m} \text{ or } h^2 = \sqrt{(4.0\text{ m})^2 + (5.7\text{ m})^2} \quad \Rightarrow h = \textbf{7.0 m.}$$

(b) $x = ?$ $h = ?$

$$\tan 58° = \frac{x}{16\text{ cm}} \quad \Rightarrow x = (16\text{ cm})\tan 58° = 25.6\text{ cm} = \textbf{26 cm}.$$

$$\cos 58° = \frac{16\text{ cm}}{h} \quad \Rightarrow h = \frac{16\text{ cm}}{\cos 58°} = \textbf{30 cm} \text{ or } h^2 = \sqrt{(16\text{ cm})^2 + (25.6\text{ cm})^2} \quad \Rightarrow h = \textbf{30 cm.}$$

2.(c) $\theta = ?$ $\alpha = ?$ $h = ?$

$$\tan \theta = \frac{6\text{ ft}}{13\text{ ft}} \quad \Rightarrow \theta = \tan^{-1}\left(\tfrac{6}{13}\right) = \textbf{25°.}$$

$$h^2 = \sqrt{(6\text{ft})^2 + (13\text{ft})^2} \quad \Rightarrow h = \textbf{14 ft.}$$

(d) $y = ?$ $x = ?$

$$\sin 21° = \frac{y}{23\text{ km}} \quad \Rightarrow y = (23\text{ km})\sin 21° = \textbf{8.3 km}.$$

$$\cos 21° = \frac{x}{23\text{ km}} \quad \Rightarrow x = (23\text{ km})\cos 21° = \textbf{22 km}.$$

Sample Problem 5

Consider the following three situations. In each case we have a 1000-kg wagon moving to the right, pulled with a 100-N force in the directions shown. How does the applied force in each case affect the motion of the wagon?

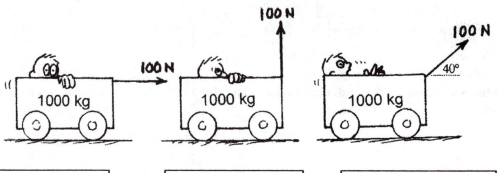

| **Case 1**: 100-N force pulling to the right | **Case 2**: 100-N force pulling straight up | **Case 3**: 100-N force pulling at an angle 40° |

Solutions: Case 1:

All of the 100-N force contributes to accelerating the cart:

$$a = \frac{F}{m} = \frac{100 \text{ N}}{1000 \text{ kg}} = 0.1 \frac{\text{m}}{\text{s}^2} ..$$

Case 2:

None of the 100-N force contributes to accelerating the cart. The force pulls up on the cart but neither raises it off the ground, nor speeds it up, nor slows it down.

Case 3:

We'd expect that the cart would accelerate somewhat, because a component of the pull is forward and "useful" (as in Case 1). As in Case 2, the upward component of force doesn't affect motion (except for reducing any friction by reducing the normal force).

The 100-N force at 40° has two components—a horizontal component (which contributes to the acceleration of the cart) and a vertical component (which, in this case, doesn't accelerate the cart). We call the horizontal direction the x-direction, and the vertical direction the y-direction.

The horizontal component, F_x, can be found from

$$\cos 40° = \frac{F_x}{100 \text{ N}} \quad \Rightarrow \quad F_x = (100 \text{ N}) \cos 40° = 77 \text{ N}.$$

So the acceleration of the cart would be $a_x = \frac{F_x}{m} = \frac{77 \text{ N}}{1000 \text{ kg}} = 0.077 \frac{\text{m}}{\text{s}^2}$.

This chapter progresses from Newton's second law to the third law, emphasizing the interactive nature of forces. When Body A exerts a force on Body B, Body B simultaneously exerts an equal force on Body A in the opposite direction. As a matter of fact, a body cannot exert a force on another body without this interaction! This chapter concludes with a treatment of vectors—first force, then velocity. The parallelogram rule is employed when adding a pair of non-parallel vectors. For more than two vectors, the parallelogram method can treat a pair at a time. More quantitatively, the Pythagorean theorem is used to find the resultant $\vec{C}$ of a pair of vectors $\vec{A}$ and $\vec{B}$ that are perpendicular to each other. From $A^2 + B^2 = C^2 \implies C = \sqrt{A^2 + B^2}$ where A, B, and C are the magnitudes of the vectors.

Sample Problem 1
With all his might, a boxer punches a sheet of paper of mass m in midair. The punch brings the paper from rest up to a speed v in time t.
(a) How much force does the boxer exert on the tissue paper?

Focus: $F = ?$

From Newton's second law $a = \dfrac{F_{net}}{m} \implies F_{net} = ma = m\dfrac{\Delta v}{\Delta t} = m\dfrac{v}{t}$.

All the terms are known quantities so the solution is complete.

(b) How much force does the tissue paper exert on the boxer?

Answer: By Newton's third law, the magnitude of the force is the same.

(c) Calculate the force of impact if the sheet of paper has mass 0.001 kg, and gains speed v of 30 m/s after an impact time of 0.050 s.

Solution: $F = m\dfrac{v}{t} = 0.001 \text{ kg}\left(\dfrac{30.0\frac{m}{s}}{0.050\text{ s}}\right) = 0.6 \text{ N}$. This is about a 2-ounce tap.

Hence the origin of the expression about not being able to fight your way out of a paper bag!

Sample Problem 2
Three blocks, A, B, and C, connected by strings of negligible mass, are pulled along a horizontal friction-free surface by a horizontal force F as shown.
Block A has a mass m, Block B has a mass $2m$, and Block C has a mass $3m$.

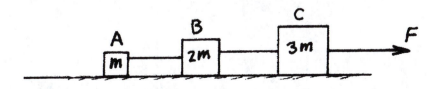

(a) Draw a force diagram showing the horizontal force(s) for the system as a whole. Find the acceleration of the system.

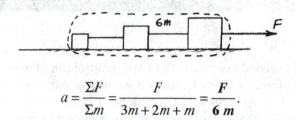

$$a = \frac{\Sigma F}{\Sigma m} = \frac{F}{3m + 2m + m} = \frac{F}{6\,m}.$$

(b) Draw a separate force diagram for Block A. How much tension is in the string that pulls Block A?

From $a = \dfrac{F_{net}}{m}$ with $T_A = F_{net\ on\ Block\ A}$ $\Rightarrow T_A = ma$.

Because the blocks are connected and slide together, each has the same acceleration.

So, $T_A = ma = m\left(\dfrac{F}{6m}\right) = \dfrac{F}{6}$.

(c) Draw a separate force diagram for Block B. How much tension is in the string that pulls backward on Block B, to the left?

The tension in the string between Blocks A and B pulls equally hard on both of them, but in opposite directions. So Block B is pulled with the same amount of tension $\dfrac{F}{6}$ toward the left.

(d) How much tension is in the string that pulls Block B to the right?

The tension accelerates the masses of both Blocks A and B.

$$T_B = (m_A + m_B)a = 3ma = 3m\left(\frac{F}{6m}\right) = \frac{F}{2}.$$

(e) How much tension is in the string that pulls backward on Block C, to the left?

The tension in the string between B and C pulls equally hard on both blocks, but in opposite directions. So the force on C from the string is 0.5 F, pulling to the left.

(f) Calculate the acceleration of the system if m is 2 kg and the pulling force is 24 N.

$$a = \frac{F}{6m} = \frac{24\ N}{6(2\ kg)} = 2\frac{m}{s^2}.$$

(g) If instead the same force F were pulling up at an angle θ from the horizontal, what would be the acceleration of the system? How would this affect the magnitudes of the tensions in the strings?

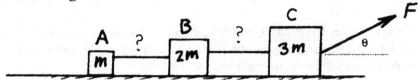

Answer: The acceleration and the string tensions would be lower. The net force horizontally would be $F\cos\theta$ so the acceleration of the system of masses would be $a = \frac{F\cos\theta}{6m}$.
Because the acceleration is less than before by a factor of $\cos\theta$, all of the tensions in the strings between blocks would also be reduced by a factor of $\cos\theta$.

• **Sample Problem 3**
A box of mass m sits on a frictionless slope at an angle θ.
(a) What is the acceleration of the box down the slope?

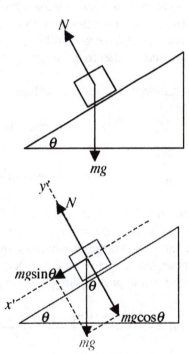

Focus: $a = ?$
$a = F_{net}/m$. Two forces act on the box—gravity (mg) and the normal force from the slope (N). The vector sum of these two forces will cause acceleration down the slope, so the net force must be down the slope. Normally we add forces in different directions by breaking them up into components, adding the components, and then calculating the resultant.

Here we'll do the same thing, except that we'll take components along a *rotated* set of axes. We'll put the rotated x'-axis pointing parallel to the slope and the rotated y'-axis perpendicular to the slope. From the geometry we see that the component of mg pointing downhill parallel to the slope is $mg\sin\theta$, and the component perpendicular to the slope is $mg\cos\theta$.

Now we can sum our forces:
$\Sigma F_{y'} = 0 \Rightarrow N - mg\cos\theta = 0$ and
$\Sigma F_{x'} = F_{net} = mg\sin\theta$.

Now we have our answer! $a = \dfrac{F_{x'}}{m} = \dfrac{mg\sin\theta}{m} = g\sin\theta$.

Let's check to see if this answer makes sense.

In the case of a vertical slope $\theta = 90°$ and $g\sin 90° = g$ (as it should, for the box would be in free fall). On a level plane $\theta = 0°$ and $g\sin 0° = 0$ (this makes sense too). And just as in free-fall, the acceleration is independent of the mass.

Look at the normal force, $N = mg\cos\theta$. When the slope is vertical $N = mg\cos 90° = 0$ (again, the box is in free-fall) and when the slope is horizontal $N = mg\cos 0° = mg$, just what we'd expect (the normal force equals the weight on a horizontal surface).

(b) Suppose that the slope is *not* frictionless, and the coefficient of kinetic friction between the box and the slope is μ_k. What is the acceleration of the box now?

Focus: a = ?
Now *three* forces act on the box—gravity (mg), the normal force from the slope (N), and the kinetic friction force (f_k) acting up and parallel to the slope.

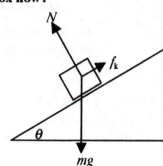

The vector sum of these three forces will cause a smaller acceleration down the slope, while the net force will still be in a direction down and parallel to the slope.

We can write our force equations:

$$\Sigma F_{y'} = 0 \quad \Rightarrow N - mg\cos\theta = 0$$

and

$$\Sigma F_{x'} = mg\sin\theta - f_k = mg\sin\theta - \mu_k N.$$

Note from $\Sigma F_{y'} = 0$ that $N = mg\cos\theta$ (and *not* $N = mg$ as it would be on a surface with no slope). This makes sense because the box presses less hard against the sloped surface than it would press against a level plane.

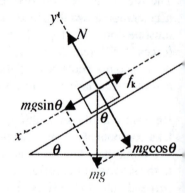

So, $a = \dfrac{\Sigma F_{x'}}{m} = \dfrac{mg\sin\theta - \mu_k N}{m} = \dfrac{mg\sin\theta - \mu_k mg\cos\theta}{m} = g\sin\theta - \mu_k g\cos\theta$

$\quad = g(\sin\theta - \mu_k\cos\theta).$

Notice the acceleration has two terms in it—the acceleration the box would have if there were no friction $(g\sin\theta)$ and a reduction in acceleration due to the kinetic friction $(\mu_k g\cos\theta)$.

(c) Calculate the acceleration of a 120-kg box on a 22° slope where the coefficient of kinetic friction between the box and the slope is 0.15.

$$a = g(\sin\theta - \mu_k\cos\theta) = 9.8\,\tfrac{m}{s^2}(\sin 22° - 0.15\cos 22°) = 2.3\,\tfrac{m}{s^2}.$$

Note that the mass of the box is irrelevant.

Newton's Third Law Problems

5-1. A car of mass m cruises along the highway at a constant velocity v. The tires push backward on the road with a force f. The reaction to this force provides the forward force on the car. Wind resistance against the car is R.

(a) What is the net force on the car?

(b) What is the acceleration of the car?

(c) A friend says that since the car is moving forward, there must be a net forward force, which means f must be greater than R, even at constant velocity. What can you say to enlighten your friend?

5-2. When Maria swims to the end of the pool she wants to increase her speed after turning around by pushing against the pool wall with force F. Her mass is m, and during her push the average water resistance is R.

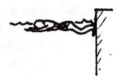

(a) What average acceleration is produced?

(b) Is it possible for Maria to push against the wall *without* the wall simultaneously pushing back on her?

5-3. A water skier of mass m is pulled at a constant velocity v by a boat of mass M. Tension in the rope held horizontally by the skier is T.

(a) What is the total resistive force by the water and air on the skier?

(b) How much upward force does the water exert on the skier?

5-4. Nellie Newton hangs at rest from a vertical rope attached to the ceiling. Consider the mass of the rope to be negligible, while Nellie's mass is m.

(a) How much force does Nellie exert on the rope?

(b) With how much force does the rope pull down on the ceiling?

(c) With how much force does the ceiling pull up on the rope?

(d) What is the net force on the rope? On Nellie?

(e) A friend says the upward and downward forces exerted on the rope make an action-reaction pair. Do you agree or disagree? Defend your answer.

5-5. • Gymnast Gracie of mass m climbs a vertical rope attached to the ceiling. The mass of the rope is negligible.

(a) What is the rope tension if she hangs motionless?

(b) What is the rope tension if she climbs at a constant rate?

(c) What is the rope tension while she accelerates up the rope with an acceleration a?

(d) What is the rope tension if she accelerates down the rope with the same magnitude of acceleration?

5-6. Block B of mass m rests on top of Block A of mass M, which rests on a horizontal table.

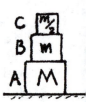

(a) Make a force diagram for each block.

(b) What is the net force on each block?

(c) What is the force (magnitude and direction) exerted by Block B on Block A?

(d) What is the force (magnitude and direction) exerted on Block B by Block A?

(e) What is the force (magnitude and direction) exerted by the table on Block A?

(f) Calculate answers to the above steps for a mass of 5.0 kg for Block B, and a mass of 7.0 kg for Block A.

5-7. A third block, Block C of mass $0.5m$, is placed on top of Block B of the previous problem. Make a free-body force diagram for each block.

(a) Make a vector diagram showing forces on the blocks.

(b) What is the net force on each block?

(c) What is the force (magnitude and direction) exerted on Block B by Block C?

(d) What is the force (magnitude and direction) exerted by Block B on Block C?

(e) What is the force (magnitude and direction) exerted by Block B on Block A?

(f) What is the force (magnitude and direction) exerted by the table on Block A?

(g) Calculate answers to the above steps for a mass of 2.5 kg for Block C, 5.0 kg for the mass of Block B, and 7.0 kg for the mass of Block A.

5-8. Pole vaulter Pablo of mass m falls from rest from a height h onto a rubber mat. The mat brings him to a halt in time t.

(a) Make a diagram of the forces that act on Pablo just as he hits the mat.

(b) What is the average force exerted on Pablo by the mat as he is brought to a halt?

(c) What is the average force on the mat as Pablo is brought to a halt?

(d) Calculate the average force on the mat if Pablo has a mass of 50.0 kg, falls from a height of 4.9 m, and is brought to a halt in 0.30 s.

5-9. David's model rocket of mass m is initially at rest. It then burns 0.050 kg of fuel in time t, ejecting it as a gas with velocity v relative to the rocket.

(a) What is the thrust of the rocket?

(b) Calculate the thrust of the rocket if its mass is 18 kg, the time during which the fuel burns is 4.0 s, and the exhaust speed of the fuel is 1200 m/s relative to the rocket.

(c) Why is David disappointed?

5-10. An astronaut of mass m floating a distance d away from her ship fires her jet pack to move toward the ship. The jet pack provides a brief thrust F toward the ship for a quick time t.

(a) What is her change in speed relative to the ship?

(b) How long does it take her to drift to her ship?

(c) How would the drift time be affected if her jet pack provided twice the thrust in the same time t?

5-11. When two identical air pucks with repelling magnets are held together on an air table and released, they move in opposite directions at the same speed v. A third identical-mass non-magnetic puck is secured to the top of one of the pucks (effectively doubling its mass) and the procedure is repeated.
   (a) How does the speed of the double-mass puck compare to the speed of the single puck?
   (b) Calculate the speed of the double-mass puck if the single puck moves away at 4 m/s.

5-12. Two astronauts initially at rest in space push on each other. The first has a mass m_1, and after the push moves backward at a speed v_1 while the second one moves at speed v_2.
   (a) What is the mass of the second astronaut?
   (b) If they push for a time t, what average force do they exert on each other?

5-13. Norman, mass M, and Karen, mass m, put on their ice skates and meet near the center of a frozen pond. While facing each other Norman pushes Karen away with force F, giving her an acceleration a. Assume that the blades of their skates remain parallel as they move apart and that friction is negligible.
   (a) What will be Norman's acceleration during the time he is in contact with Karen? What is the direction of his acceleration?
   (b) Calculate Norman's acceleration if his mass is 66 kg, Karen's mass is 44 kg, and she accelerates at 2.4 m/s^2.

5-14. • Mady with mass m and Sam with mass M stand on frictionless ice x meters apart. Sam pulls on a rope that connects him to Mady, giving her an acceleration a toward him.
   (a) What is Sam's acceleration?
   (b) Calculate Sam's acceleration given that his mass is 66 kg, Mady's mass is 55 kg, and her acceleration toward him is 0.90 m/s^2.
   (c) Calculate where Mady and Sam will meet if they are initially 10 m apart.

5-15. In a tug-of-war on a quite slippery floor Jose of mass M pulls on the rope and imparts an acceleration a to Juanita of mass m.
   (a) What is the tension in the rope?
   (b) Calculate the tension if Jose's mass is 80 kg, Juanita's mass is 60 kg, and she accelerates toward Jose at 0.10 m/s^2.

5-16. Ramon of mass m stands on a scale in an elevator of mass M that is momentarily at rest.
   (a) What is the reading of the scale?
   (b) What is the reading when the elevator moves upward at constant velocity v?
   (c) What is the reading when moving upward with an acceleration a?
   (d) Calculate readings for Ramon whose mass is 70.0 kg, the constant velocity in (b) is 1.5 m/s, and the constant acceleration in (c) is 2.0 m/s^2.

5-17. Maria Moonglow stands on a bathroom scale in the elevator of the physics building. When at rest, the scale shows a weight W. As the elevator starts to moves up, the scale reading is 1.2 W.
(a) Find the acceleration of the elevator.
(b) Maria who normally weighs 550 N. What is her weight as recorded by the scale in the accelerating elevator?
(c) Calculate her "elevator weight" when the elevator is moving downward with the same magnitude of acceleration that you calculated in step (a) above.

5-18. Consider the tension in the strands of cable that move an elevator upward in a skyscraper of height y. The elevator with its load of people has mass M, and moves upward at constant speed v.
(a) What is the combined tension in the supporting cables?
(b) Calculate the tension in the cables that moves the elevator of mass 2000.0 kg upward a distance 292 m at a constant speed 10 m/s.
(c) Why would this problem have been appropriate for Chapter 2?

5-19. A car of mass m and a truck of mass $2m$ traveling at the same speed v have a head-on collision. During the collision the truck undergoes an average deceleration a.
(a) What deceleration does the car undergo?
(b) Calculate the car's deceleration during the collision if its mass is 958 kg, initial speed is 15 m/s, and impact time is 0.20 s.
(c) Calculate the average force on the car during collision.
(d) If you simulate this with smaller carts in lab, each with a force sensor at its "front bumper," how will the force readings for both compare during the collision?

5-20. A giant frog of mass m is placed on a skateboard of mass M at rest. When the frog jumps in a forward direction from the skateboard, the skateboard rolls freely in the opposite direction at speed V.
(a) What was the frog's horizontal jumping speed?
(b) Calculate the frog's horizontal jumping speed if its mass is 2.0 kg, the skateboard's mass is 3.0 kg, and the skateboard recoils at 1.4 m/s immediately after the jump.
(c) A force of friction is needed for the frog to propel horizontally from the skateboard (a slippery surface won't do). Can the skateboard surface provide a friction force on the frog without an equal and opposite force of friction on the skateboard? Defend your answer.

5-21. Three blocks of equal mass m, each connected by a string of negligible mass, are pulled along a frictionless surface by a horizontal force F as shown in the sketch.
(a) What is the acceleration of the system?
(b) What is the tension in each string? (Hint: Draw a force diagram for the system, and for each block separately.)
(c) Calculate the acceleration and the string tensions if the mass of each block is 10 kg and the force F is 60 N.

5-22. Phil exerts a force F on two crates, one in front of the other, as shown. Crate A has a mass m, while Crate B has a smaller mass, $0.5m$. The crates are mounted on tiny rollers and move with negligible friction.

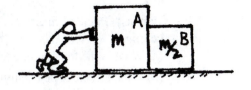

(a) Make a force diagram for the system consisting of Crate A + Crate B.

(b) What is the acceleration of the two-crate system?

(c) Make a force diagram for Crate B. How great a force acts on Crate B while Phil continues pushing?

(d) Make a force diagram for Crate A. How many horizontal forces act on it, and what net force acts on it while Phil continues pushing?

(e) What would be different in this problem if Crates A and B were interchanged?

(f) A friend says that Crate A accelerates because Phil's hands push it, but Crate B has no force on it so just rides along with Crate A. What physics is your friend missing?

5-23. Christopher is at the rear of a bus that travels along a straight level road at velocity v_1. He tosses a ball at velocity v_2 relative to the bus toward the front of the bus.

(a) What is the velocity of the ball relative to Earth?

(b) Suppose he instead throws the ball from the front of the bus toward the rear. What is the velocity of the ball relative to Earth?

5-24. Two scales are used in a classroom demonstration to suspend a 10-N weight.

(a) Explain how each of the scales can register 10 N. That is, how can 10 N + 10 N = 10 N?

(b) Show your answer with a force diagram.

5-25. Consider the combination of two forces, F_1 and F_2.

(a) What is the largest possible force that can result from the combination? Draw a sketch to illustrate your answer.

(b) What is the smallest possible force that can result from the combination? Draw a sketch to illustrate your answer.

(c) Give numerical values to the above if $F_1 = 12$ N and $F_2 = 18$ N.

5-26. A baseball of mass m is batted at velocity v at an angle θ with the horizontal.

(a) Find the horizontal and vertical components of the ball's velocity.

(b) Calculate the horizontal and vertical components for an initial velocity of 45 m/s at angle 37° to the horizontal.

5-27. Gymnast Gracie of mass m is suspended by a pair of vertical ropes attached to the ceiling.
   (a) What is the tension in each rope?
   (b) What are the rope tensions if they comprise a V-shape, each at an angle θ with the ceiling?
   (c) Suppose the gymnast has a mass of 55 kg. Calculate answers for tension in the pair of vertical ropes, and for when they are 53° to the ceiling.

5-28. Bronco Brown mowing his lawn applies a force F along the handle of the lawnmower as shown. Bronco moves the mower at constant velocity.
   (a) What resistance force acts horizontally on the mower?
   (b) Why is "constant velocity" the key to solving this problem?
   (c) Calculate the resistant force if Bronco applies 208 N to the handle, the mower moves at 1.2 m/s, and the angle of the handle is 45°.

5-29. The weight W of a box on an inclined plane can be resolved into two vector components, one parallel to the plane and the other perpendicular to it.
   (a) At what angle of the plane relative to the horizontal would the components be equal?
   (b) At what angle would the perpendicular component equal W?
   (c) At what angle would the parallel component equal W?
   (d) Assuming negligible friction, what is the acceleration of the box in all four cases?

5-30. Luigi wishes to suspend a lantern of mass m between his store and a telephone pole. The cable attached to his store makes an angle θ to the wall as shown. The cable to the telephone pole is horizontal.
   (a) What is the tension in the cable of angle θ?
   (b) What is the tension in the horizontal cable?
   (c) Calculate the two tensions for a 12-kg lantern and $\theta = 40°$.

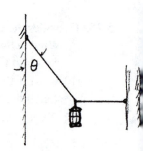

5-31. • A street lamp is suspended by two cables, one at angle θ_1 and the other at angle θ_2 as shown.
   (a) Find the tension in each cable for a lamp of mass m.
   (b) Calculate the two tensions when the mass of the lamp is 15 kg, θ_1 is 35°, and θ_2 is 55°.

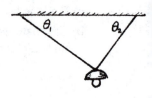

5-32. Snowboarder Sherry of mass m is pulled up a slope at constant velocity v by a tow bar. The force of the tow bar is parallel to the slope, which is inclined at θ with respect to the horizontal. The coefficient of kinetic friction between the snowboard and the snow is μ_k.
(a) Find the force that the tow bar exerts on Sherry.
(b) Calculate the force for a 45.0-kg snowboarder moving at 3.0 m/s on a 20°-slope if the coefficient of kinetic friction is 0.110.

5-33. Larry pulls a log of mass m up a ramp at constant speed by a rope that is parallel to the ramp surface. The ramp is inclined at θ with respect to the horizontal, and the coefficient of kinetic friction is μ_k.
(a) Find the tension in the rope.
(b) Find the tension in the rope if the log undergoes a steady acceleration a.
(c) Calculate the tensions for constant speed, and for acceleration 0.4 m/s² if the mass of the log is 195 kg and the ramp is inclined at 37° with respect to the horizontal. Assume $\mu_k = 0.30$.

5-34. • Tammy pulls a rope at an angle θ above the horizontal on a box with mass m. The coefficients of friction between the box and the surface are known, both μ_s and μ_k.
(a) What tension in the rope is required to get the box to start moving?
(b) A rope pulls at an angle of 20° above the horizontal on a 23-kg box. The coefficients of friction between the box and the surface are $\mu_s = 0.32$ and $\mu_k = 0.25$. Calculate the required tension in the rope to get the box to start moving.
(c) Calculate the acceleration of the box if she continues pulling with the force you found above.

Show-That Problems for Newton's Third Law

5-35. Amanda looks at a 1.00-kg bag of jellybeans resting on a table.
Show that the amount of force that the table exerts on the bag of jellybeans is 9.8 N.
How does this compare with the force the bag of jellybeans exerts on the table?

5-36. Suppose the table in the previous problem was somehow accelerated upward at one-half g.
Show that the force of the table on the jellybeans would be 14.7 N.

5-37. Suppose the table in the previous problem was accelerated downward at one-half g.
Show that the force of the table on the jellybeans would be 4.9 N.

5-38. When 60-kg Victor on rollerblades pushes against a wall with a force of 30 N he recoils.
Show that the recoil acceleration is 0.5 m/s^2.

5-39. A 7.00-kg bowling ball moving at 8.00 m/s strikes a 1.00-kg bowling pin and slows to 7.00 m/s in 0.040 s.
Show that the force of impact is 175 N.

5-40. Two people, one with three times the mass of the other, attempt a tug-of-war on frictionless ice.
Show that the heavier person will gain a speed one-third that of the lighter person.

5-41. Carts A and B are connected by a compressed spring on an air table. Cart A has a mass of 0.25 kg and Cart B has a mass of 0.75 kg. The spring is then released.
Show that Cart A moves at three times the speed of Cart B.

5-42. Consider an airplane with an airspeed of 125 km/h flying with its nose pointing due east. It then encounters a north wind blowing at 48 km/h.
Show that its ground speed is 134 km/h.

5-43. You take your 32-kg nephew to play on the swings at the park. You pull him 20° from the vertical and hold him stationary before letting him go.
Show that the combined tensions in the swing ropes equal 334 N. Then show that the horizontal force you exert in holding him stationary is 114 N.

5-44. Refer to the previous problem.
Show that the tension in each of the two ropes is $mg/(2\cos\theta) = 167$ N.

5-45. • Consider a force F directed at an angle θ below the horizontal on a box of mass m that sits on a level surface. Coefficient of static friction is μ_s.
Show that the force required to get the box to move must be at least $\dfrac{\mu_s mg}{\cos\theta - \mu_s \sin\theta}$. (Hint: Note that the normal force is greater than mg.)

6 *Momentum*

Momentum, inertia in motion, is a vector quantity equal to mass × velocity. In the absence of a net external force, the momentum of a body (or system of bodies) is conserved. If a net force *does* act on the system for a time interval t, then $F_{net}t = \Delta(mv)$. The quantity $F_{net}t$, the product of the force acting on the body (or system of bodies) multiplied by the time interval over which the force acts, is called the *impulse*. The equation tells us that an impulse acting on a body causes an equivalent change in the momentum of that body.

Usually a change in momentum involves an object of fixed mass changing velocity, in which case $\Delta(mv) = m\Delta v = m(v_f - v_0)$. The v's in this equation stand for velocities. As you will see in the first sample problem, it is not only the magnitude (speed) of the body that matters—the direction matters as well.

Problems for this chapter include the impulse-momentum relationship, momentum conservation, and near the end, a bit on vectors.

Sample Problem 1
A horizontal stream of water is directed against the *locked* blade of a Pelton wheel, as shown. The water strikes the blade at velocity v and bounces in the opposite direction at the same speed. In time t a mass m of water rebounds.

(a) What magnitude of force acts on the blade and the water?

Focus: $F = ?$

Consider first the force on the water. From $Ft = \Delta mv \Rightarrow F = \dfrac{\Delta mv}{t} = \dfrac{m\Delta v}{t}$. Since the water strikes with initial velocity v and rebounds with velocity $-v$, its velocity changes by an amount $2v$. So the magnitude of the force on the water is $F = \dfrac{m(2v)}{t} = \dfrac{2mv}{t}$. From Newton's third law, the force on the blade has the same magnitude (but is opposite in direction).

(b) Calculate the force on the blade if 22 kg of water strikes the blade each second at a speed of 25 m/s and rebounds at the same speed.

$$F = \frac{2mv}{t} = \frac{2(22\,\text{kg})(25\,\tfrac{\text{m}}{\text{s}})}{1\text{s}} = 1100\,\text{N}.$$

(c) Will this force be more, less, or the same, for the case of a moving blade?

Answer: The relative speed of impact of the water on the blade will be less when the blade moves in the same direction as the water, so **the force on the blade will be reduced**. To exaggerate: If the blade moved as fast as the water, there would be no force of impact.

Sample Problem 2

A giant frog of mass m is placed on a skateboard of mass M, initially at rest. When the frog jumps horizontally in a forward direction from the skateboard, the skateboard rolls backward at speed V.

(a) What was the frog's jumping speed?

Focus: $v = ?$

No horizontal external forces act on the (frog + skateboard) system, so the momentum of the system is conserved. Since the initial momentum of the (frog + skateboard) system is zero, the final momentum of the system must be zero as well. That means the momentum of the jumping frog is equal and opposite to the momentum of the recoiling skateboard. The magnitudes of momentum are the same for both.

$$\text{From } m_{frog}v_{frog} = M_{board}V_{board} \Rightarrow v_{frog} = \frac{M_{board}V_{board}}{m_{frog}} = \frac{M(V)}{m} = \frac{MV}{m}.$$

(b) Calculate the frog's horizontal jumping speed if its mass is 2.0 kg, and the skateboard of mass 3.0 kg recoils at 1.4 m/s immediately after the jump.

$$v_{frog} = \frac{MV}{m} = \frac{(3.0 \text{ kg})(1.4 \text{ m/s})}{2.0 \text{ kg}} = 2.1 \text{ m/s}.$$

Note that this is a problem from Chapter 5! Conservation of momentum provides a quicker route to the solution.

Sample Problem 3

A giant frog of mass m drops straight down from a tree branch onto a skateboard of mass M that is moving horizontally beneath it at speed V.

(a) Find the speed of the skateboard after the frog lands on it.

Focus: $v = ?$

Again, we employ the law of conservation of momentum.

Since no external forces act in the horizontal direction on the skateboard + frog system, the horizontal momentum of the skateboard before the frog lands is the same as the momentum of the skateboard + frog after the frog lands.

$$\text{From } MV_{before} = (M + m)v_{after} \Rightarrow v_{after} = \frac{M}{M + m}V_{before}.$$

(b) Calculate the speed of the skateboard after the frog lands on it. The skateboard was initially moving 4.0 m/s. The mass of the frog is 2.0 kg and the mass of the skateboard is 3.0 kg.

$$\textit{Solution: } v_{skateboard + frog} = \frac{M}{M + m}V_{before} = \frac{3.0 \text{ kg}}{3.0 \text{ kg} + 2.0 \text{ kg}}(4.0 \text{ m/s}) = 2.4 \text{ m/s}.$$

(c) When the frog lands, a force of friction keeps him on the skateboard (for a slippery skateboard the frog would end up on the ground). Can the skateboard surface provide a friction force on the frog without an equal and opposite force of friction on the skateboard? Defend your answer.

Answer: **No.** In accord with Newton's third law, the skateboard can supply a force (friction or otherwise) on the frog only if the frog supplies an equal and opposite force on the board.

(d) What is the impulse that slows the skateboard?

> *Answer*: The impulse that reduces the momentum of the skateboard is $-ft$, the friction force of the frog's feet acting backwards on the skateboard × time. The equal and opposite friction force of the skateboard on the frog's feet × time provides the forward impulse on the frog to bring it up to speed.

Note that the net horizontal momentum of the (frog + skateboard) system is the same before and after the frog lands—because no *external* friction forces act (such as between the ground and the skateboard).

Momentum Problems

6-1. An average net force F applied for time t is required to accelerate an elevator of mass m from rest up to a steady speed v.
(a) What impulse was applied?
(b) What was the momentum of the elevator when it reached a steady speed?

6-2. Lonnie's truck of mass m moves at speed v into a headwind of speed V. The temperature of the road surface is T.
(a) What is the momentum of the truck?
(b) Calculate the momentum of Lonnie's 3500-kg truck moving at 18 m/s when it drives into a 5 m/s headwind, and when the temperature of the road is 24°C.
(c) How does this problem show the importance of letting equations guide your thinking?

6-3. A block of ice, mass m, slides from rest down an inclined plane. At the bottom of the incline it slides onto the floor at speed v.
(a) What momentum does the block gain in sliding down the incline?
(b) Find the momentum gained by a 25-kg block of ice that starts from rest and slides off the end of an inclined plane at a speed of 6.0 m/s.

6-4. A bullet of mass m exits a gun barrel at a speed v.
(a) What impulse acted on the bullet?
(b) Calculate the impulse that acted on a 9.6-gram bullet that leaves the gun barrel at a speed of 280 m/s.
(c) Impulse = force × time, neither or which is given information in this problem. Why does this absence of data not affect your calculation?

6-5. A soccer player kicks a ball initially at rest with an average force F. The player's foot remains in contact with the ball for a brief time t.
(a) What is the impulse that acts on the ball?
(b) What is the change in momentum of the ball?
(c) The ball has a mass of 0.42 kg, the average force of the kick is 1350 N and time the foot acts on the ball is 0.0080 s. Calculate the speed of the ball after the kick.

6-6. In the hammer throw contest at a track and field event, Dean whirls a hammer of mass m in a circle and releases it at speed v.
(a) What impulse acted on the hammer?
(b) Calculate the impulse if the mass of the hammer is 7.3 kg and it is released at a speed of 28 m/s.

6-7. A diver of mass m atop a high cliff dives straight down into the water below. Just before striking the water her speed is v_1. After a brief time t in the water her speed is reduced to v_2.
   (a) What is the average net force exerted on the diver that reduces her speed?
   (b) Calculate the net force that slows the 48.0-kg diver when her speed just before hitting the water is 9.9 m/s and reduces it to 0.5 m/s after 0.60 s.

6-8. When an abrupt average force F is exerted on a system, the system's momentum increases by mv.
   (a) What is the time during which the force acts?
   (b) If a different force acts on the system for twice as much time and produces the same change in momentum, how will the "new" average force compare with the original force?

6-9. A baseball is tossed straight upward with an initial momentum mv.
   (a) If air drag is negligible, what will be the ball's momentum when caught at the same elevation from which it was thrown?
   (b) If air drag is not negligible, does the baseball's momentum just before it is caught differ from its initial momentum? How does it differ?

6-10. An unfortunate bird with momentum mv flies into a glass window, which brings the bird to an abrupt stop.
   (a) What is the magnitude of the impulse that stopped the bird?
   (b) The bird recovers. If it had instead bounced from the window, would its recovery be more likely or less likely? Defend your answer.

6-11. Diane spikes a volleyball of mass m so that its incoming speed v is changed to an outgoing speed V in the opposite direction.
   (a) What is the magnitude of the impulse that acts on the volleyball?
   (b) Calculate the impulse when the mass of the volleyball is 0.25 kg, its incoming speed is 4 m/s, and outgoing speed in the opposite direction is 20 m/s.

6-12. A golf ball of mass m, initially at rest, is given a speed v when the time of contact between the club and ball is t.
   (a) What is the average force on the ball?
   (b) Calculate the average force for the 0.045-kg golf ball that gains a speed of 24.0 m/s when the time of contact between club and ball is 0.0020 s.

6-13. Pedro throws a baseball (mass m) at a wall at a speed V. It rebounds from the wall with a speed v.
   (a) If the contact time of the ball with the wall is t, what is the magnitude and direction of the average force that the ball exerts on the wall?
   (b) Calculate the average force on the wall from a ball of mass 0.15 kg thrown at a speed of 35 m/s that leaves the wall at 22 m/s after a 0.0030-s impact.
   (c) How does the magnitude of the force of the ball on the wall compare with the magnitude of the force the wall exerts on the ball?

6-14. In 0.010 s, 0.0050 kg of exhaust gas is ejected at speed v from a model rocket of mass m sitting on its launch pad.
   (a) What is the thrust on the rocket?
   (b) Calculate the thrust on the rocket if its mass is 1.8 kg and the exhaust speed of the fuel is 110 m/s relative to the rocket.

6-15. When hit with a bat, a baseball of mass m changes velocity from v_1 toward the bat to v_2 away from the bat in time t.
  (a) How much impulse acts on the ball?
  (b) Calculate the impulse acting on a 0.15-kg baseball that changes speed from 45.0 m/s to 55.0 m/s in the opposite direction.
  (c) Calculate the average force on the baseball for a ball-bat contact time of 0.0020 s.
  (d) How does the average force on the ball compare with the average force on the bat?

6-16. A jet engine gets its thrust by taking in air, heating and compressing it, and then ejecting it at a high speed. A particular engine takes in a mass of air m per time t at speed v, and ejects it at speed $10v$.
  (a) Find the thrust of the engine.
  (b) Calculate the thrust of the engine if it takes in 20 kg of air per second at 100 m/s and ejects it at 1000 m/s.

6-17. A chunk of ice of mass m breaks loose from a suspension bridge and falls for time t.
  (a) Neglecting air resistance, what will be its momentum when it hits the water below?
  (b) What would be its momentum falling from a higher bridge where the falling time is $2t$?

6-18. Manuel drops a water balloon of mass m from rest atop the roof of a building of height h.
  (a) What is its momentum when the balloon hits the street below?
  (b) If the water balloon has a mass of 1.5 kg and the building is 12 m high, calculate the momentum of the balloon when it hits the street?
  (c) Why can't the force of impact be found using only the information given above?

6-19. A pole-vaulter of mass m falls from rest from a height h onto a rubber mat. After hitting the mat our vaulter comes to rest in time t.
  (a) Make a diagram of the forces that act on the pole vaulter just as he hits the mat.
  (b) What is the average force exerted on the vaulter by the mat as he is brought to a halt?
  (c) Calculate the average force exerted by the mat on the pole vaulter if he has a mass of 55 kg, falls from a height of 4.9 m, and comes to rest 0.30 s after hitting the mat.

6-20. An astronaut of mass m is floating a distance d away from a stationary ship in outer space. She fires her jet pack to move towards the ship. The jet pack provides a 0.5-second thrust F towards the ship.
  (a) What gain in speed does she experience?
  (b) Assume the distance traveled while accelerating is negligible compared with her distance from the ship. How long does it take her to return?
  (c) What is the time of her return if her jet pack provides twice the thrust?

6-21. • A steel ball of mass m is dropped from a height h above the flat, horizontal surface of a massive block of iron. It rebounds to a height $0.80h$ above the iron.
  (a) What impulse acts on the ball during impact?
  (b) If the ball-iron contact time is 0.0018 s, find the average impact force on the ball.
  (c) Calculate the average force if the mass of the ball is 0.40 kg and height h is 1.0 m.

6-22. A wad of putty of mass m drops vertically onto a cart of mass M that rolls horizontally at a speed v.
  (a) What is the speed of the cart after the putty sticks to it?
  (b) Calculate the speed of the cart if the putty's mass is 5.0 kg, the mass of the cart is 10.0 kg, and the initial speed of the cart is 5.0 m/s.

6-23. A lump of clay of mass m_1 and velocity v_1 catches up with and bumps into a slower lump of clay of mass m_2 and velocity v_2 heading in the same direction. They share a common velocity after they stick together.
(a) What is this common velocity?
(b) Calculate the final velocity of the stuck-together lumps when a lump of mass 2.2 kg moving at 3.2 m/s catches up with and sticks to a 2.8-kg lump moving at 1.2 m/s.

6-24. Lynda of mass m_1 is on roller skates and moves at speed v_1 when she crashes into and hugs Duncan (mass m_2) initially at rest and also on skates.
(a) With what velocity do they skate off together into the sunset?
(b) Calculate their sunset velocity if Duncan's mass is 64.0 kg, Lynda's mass is 45.0 kg and she initially moves toward Duncan at speed 4.5 m/s.

6-25. A blob of putty of mass m has a head-on collision with another blob of putty of mass $2m$ moving toward it. After the collision the combined putty blobs don't move.
(a) What can you conclude about the relative speeds of the blobs before collision?
(b) If the smaller blob was moving at a speed of 2 m/s before the collision, what was the speed of the other blob before the collision?

6-26. A dart of mass m moves horizontally at speed v and strikes and sticks to a wooden block of mass M initially at rest on a horizontal friction-free surface.
(a) What is the resulting speed of the dart-block system?
(b) If the dart's mass is 1.0 kg, its initial speed is 8.0 m/s, and the mass of the wooden block is 12.0 kg, calculate the "after-collision" speed for the dart-block system.

6-27. Duncan's car of mass m moving at speed v on an icy road collides with a stationary truck of mass M.
(a) What is the speed of the interlocked vehicles as they slide along the icy road?
(b) Calculate the speed of interlocked vehicles if the car's mass is 1500 kg, the truck's mass is 3200 kg, and the initial speed of the car is 22 m/s.

6-28. Two identical air pucks with repelling magnets are held together on an air table. When they are released the pucks move away from each other at equal speeds. Then a third identical non-magnetic puck is secured to the top of one of the pucks (resulting in twice the mass) and the procedure is repeated.
(a) How will the speed of the double-puck compare to the speed of the single puck?
(b) Calculate the speed of the double-mass puck if the single puck moves away at 4 m/s.

6-29. Suppose a car of mass M moving at speed V has a head-on collision with a smaller car of mass m moving at speed v toward it.
(a) What is the speed of the coupled cars immediately after collision?
(b) Calculate the speed for the coupled cars if the mass of the larger car is 1500 kg, its initial speed is 18 m/s, and the mass of the smaller car is 1100 kg with initial speed 29 m/s in the opposite direction.

6-30. When a bullet of mass m strikes and is imbedded in a wooden block of mass M, the bullet-block system moves at speed V in the direction of the moving bullet.
(a) With this information, find the initial speed of the bullet.
(b) Calculate the bullet's initial speed if its mass is 0.050 kg, the mass of the block is 5.0 kg, and the block-bullet system moves 4.0 m/s in the direction of the moving bullet.

6-31. Sumo wrestler Akebono of mass m collides head on with Konishiki of mass M. Konishiki moves with a speed of V toward Akebono and the collision brings their combined speed to zero.
(a) How fast must Akebono have been moving before the collision?
(b) For the speed to be zero when the wrestlers combine, how fast does the 227-kg Akebono move if 267-kg Konishiki moves at 2.0 m/s toward Akebono?

6-32. Right after two astronauts initially at rest in space push on each other, the first of mass m_1 moves away at speed v_1 while the second one moves in the opposite direction at v_2.
(a) What is the mass of the second astronaut?
(b) If they push on each other for a total time t, what average force does each astronaut exert on the other?

6-33. Carmelita loosely holds and fires a rifle of mass M that imparts a momentum mv to the bullet as it travels through the barrel.
(a) What is the recoil velocity of the gun?
(b) How would recoil velocity be affected if Carmelita holds the rifle tightly? Explain.

6-34. A launcher of mass M resting on a horizontal, frictionless surface fires a projectile of mass m horizontally at speed v relative to the surface.
(a) What is the recoil speed of the launcher?
(b) Calculate the recoil speed for a 2400-kg launcher that fires a 38-kg projectile horizontally at 520 m/s.

6-35. A quarterback of mass m moving north at speed v is tackled by an opponent of mass M running south at speed V.
(a) What is their combined speed when they tangle? (Ignore friction with the ground.)
(b) A 89-kg quarterback moving north with a speed of 3.5 m/s is tackled by a 110-kg opponent running south at 5.0 m/s. Calculate their combined speed after they become tangled, ignoring their grunts and friction with the ground.

6-36. A hockey puck moving at a speed V_1 collides head on with a second identical puck moving toward it at speed V_2. After the collision, the first puck slows down to speed v_1 without changing direction.
(a) After the collision what is the speed v_2 of the second puck?
(b) Calculate the speed v_2 of the second puck when the first puck had an initial speed of 18 m/s that was changed to 2.0 m/s by the collision, and the initial speed of the second puck was 12.0 m/s. Both pucks have a mass of 0.16 kg.
(c) Does your answer change if the masses of both pucks are doubled?

6-37. A baseball of mass m leaves the bat at a speed v and makes an angle θ with the horizontal.
(a) Find the horizontal and vertical components of the ball's momentum.
(b) Calculate the horizontal and vertical components of momentum if the ball's mass is 0.15 kg, its speed is 45 m/s, and the angle θ is 37°.

6-38. A golf ball of mass m is given a speed v at an angle θ with the horizontal.
(a) Find the horizontal and vertical components of the ball's momentum.
(b) Calculate the horizontal and vertical components of momentum if the ball's mass is 0.050 kg, speed is 35 m/s, and angle θ is 53°.

6-39. • A blue convertible of momentum mv traveling north collides with a red jeep having the same magnitude of momentum, but traveling west.

    (a) What is the magnitude and direction of the resulting momentum of the vehicles if they stick?

    (b) Calculate the resulting momentum if the mass of each vehicle is 1800 kg and their initial speeds were both 12 m/s.

6-40. • A truck of mass M traveling east at a speed v collides with a car of mass m moving north at equal speed v. After collision both vehicles remain tangled together.

    (a) With what speed and in what direction does the wreckage move?

    (b) If the coefficient of kinetic friction between the wreck and the road is μ_k how far does the wreck slide along the road?

Show-That Problems

6-41. A force of 10.0 N acts for 0.010 s on an object.

    Show that the change in momentum of the object will be 0.10 kg·m/s.

6-42. A 20.0-kg mass moving at a speed of 3.0 m/s is stopped by a constant force of 15.0 N. Show that the stopping time required is 4.0 s.

6-43. A braking force is needed to bring a 1200-kg car moving at 25 m/s to rest in 20.0 s. Show that the braking force is 1500 N.

6-44. A 50-gram egg is thrown at 4 m/s at a bed sheet and is brought to rest in 0.2 s. Show that the average force of egg impact is 1 N.

6-45. An 84.0-kg person in a car moving at 24.0 m/s is brought to rest in 1.20 s by an air bag. Show that the approximate force exerted by the air bag on the person is 1680 N.

6-46. A golf ball of mass 0.045 kg traveling horizontally at 28 m/s hits a brick wall and bounces back at the same speed after a contact time of 0.040 s. Show that the force of impact is 63 N.

6-47. A 115-kg astronaut throws an 18-kg tool kit at 4.6 m/s away from a spacecraft in deep space.

    Show that the astronaut will recoil at a speed of 0.72 m/s.

6-48. Sam (90.0 kg) and Mady (60.0 kg) stand 10.0 m apart on ice and have a tug-of-war. Big Josh sits on the side and watches. As Sam and Mady pull, Big Josh sees Sam move 2.0 m along the ice toward Mady.

    Show that Big Josh sees Mady move 3.0 m along the ice toward Sam.

6-49. A 2.0-kg object traveling 10.0 m/s north has a perfectly elastic collision with a 5.0-kg object traveling 4.0 m/s south.

    Show that the combined momentum after the collision is zero.

6-50. A car of mass 1200 kg travels north at 18 m/s and has an inelastic collision with a SUV of mass 2400 kg traveling east at 12 m/s.

    Show that the entangled wreck slides at 53° from a northwardly direction.

Energy

We move from the vector quantity momentum to the scalar quantity energy. Mechanical energy involves both kinetic energy (KE) and potential energy (PE). Work (W) done on a system changes the energy of the system, either by lifting something and increasing its PE, or by changing its speed and therefore its KE. The rate at which work is performed is called Power (P). The energy of stretched and compressed springs is treated later in the problem set.

Ask a professional physicist to solve a typical mechanics problem and the most likely approach will be energy conservation. Many of the problems we've previously treated, whether in kinematics or dynamics, are nicely solved by either the work-energy theorem or conservation of energy. Applying energy concepts to mechanics problems often shortens laborious solutions.

Sample Problem 1

Manuel drops a water balloon of mass m from rest atop the roof of a building. The balloon takes time t to hit the ground below.
(a) What is its KE just before it hits?

Focus: KE =?

Energy conservation applies to this problem. PE at the top of the building becomes KE at the bottom. $KE = mgh$. But we're not given h.

From kinematics $h = \frac{1}{2}gt^2$ so $KE = mg\left(\frac{1}{2}gt^2\right) = \frac{1}{2}mg^2t^2$. Or from

$$v_f = v_0 + at \implies v_{\text{at bottom}} = gt \implies KE_{\text{at bottom}} = \frac{1}{2}mv^2_{\text{at bottom}} = \frac{1}{2}m(gt)^2 = \frac{1}{2}mg^2t^2.$$

(b) If the water balloon has a mass of 1.2 kg and the dropping time from rest is 2.0 s, what will be its kinetic energy when it makes impact with the ground below?

$$KE = \frac{1}{2}mg^2t^2 = \frac{1}{2}(1.2\text{kg})\left(9.8\frac{\text{m}}{\text{s}^2}\right)^2 (2.0\text{s})^2 = 230\,\text{kg}\frac{\text{m}^2}{\text{s}^4}\text{s}^2 = \mathbf{230\,J}\,.$$

(c) Why can't the force of impact be found with the information given?

Answer: To find the force of impact we'd have to know either the impact time or the impact distance. We don't know either one.

Sample Problem 2

Janet is cruising along a level road at speed v and slams on the car's old-fashioned non-antilock brakes and slides to a stop. The coefficient of kinetic friction is μ_k.
(a) How far does Janet's car slide?

Focus: $d = ?$

We consider the work-energy theorem (much more central than the kinematics equations!). It is

$$W = \Delta KE$$
$$\text{or, } Fd = \Delta\left(\frac{1}{2}mv^2\right).$$

Here we have work done by the force of friction, which acts in a direction opposite to the car's direction of motion to slow the car. The distance of slide is d, the mass of the car is m, and the car's initial speed is v. In this problem the final speed of the car will be zero, so the amount of change in KE is equal to the initial KE at speed v. We're looking for distance, so we write the equation

$$d = \frac{\Delta(\frac{1}{2}mv^2)}{F}.$$

If all the terms are known, this would be a simple one-step problem. But all the terms in the equation are not known values. Progress to a solution is dictated by finding the unknown terms. F is the force of friction, f. So our equation becomes

$$d = \frac{\Delta(\frac{1}{2}mv^2)}{F} = \frac{\frac{1}{2}mv^2}{f}.$$

Now $f = \mu_k N$, where $N = mg$ since the road is horizontal. So, keeping it all neatly on a single horizontal line to save vertical space, the solution is:

$$d = \frac{\Delta(\frac{1}{2}mv^2)}{F} = \frac{\frac{1}{2}mv^2}{f} = \frac{\frac{1}{2}mv^2}{\mu_k N} = \frac{\frac{1}{2}mv^2}{\mu_k mg} = \frac{v^2}{2\mu_k g}.$$

Note how the terms in the equation dictate subsequent steps. Again, equations are more than recipes for plugging in values; they can also be guides to thinking.

(b) Does the solution make sense? And what does it tell us?

Answer: The solution tells us the stopping distance is proportional to speed squared, which is consistent with KE—a car going twice as fast will have four times as much KE, and so will require four times the distance for the same friction force to bring it to a stop. It also tells us that if μ_k or g is greater the stopping distance will be less because the force of friction will be greater. That makes sense. And we see the mass cancels in the equations, which tells us that the mass of the car doesn't matter. All cars skidding with the same initial speed, with equal coefficients of kinetic friction, will skid the same distance. And as for units, note that $\frac{v^2}{2\mu_k g}$ has units $\frac{m^2}{s^2} / \frac{m}{s^2} = m$, which is a distance, as it should be.

Good physics!

Sample Problem 3

Phil pushes Mala across the snow on her sled. He pushes at a downward angle θ as shown.

(a) How much work does Phil do on the sled by his applied force F pushing the sled a horizontal distance d?

Focus: $d = ?$

Note that the solution is NOT simply Fd, because F is not in the direction of the sled's motion. When an applied force is at an angle to the direction of motion, work is calculated by multiplying the *component* of the force parallel to the direction of motion, by the distance moved.

In this case,

$$W = F_x d = (F\cos\theta)d, \text{ or } Fd\cos\theta.$$

(b) Calculate the speed of the 62-kg sled, initially at rest, after being pushed a distance of 6.5 m by a 35-N push directed at an angle of 25° below the horizontal. Assume negligible friction.

Focus: $v = ?$

If this problem were given in Chapters 4 or 5, before we learned about the work-energy theorem, our solution would likely go thusly:

From $v_f^2 - v_0^2 = 2ad$, with $v_0 = 0 \Rightarrow v = \sqrt{2ad}$. From $a_x = \dfrac{F_x}{m} \Rightarrow a = \dfrac{F\cos\theta}{m}$ so $v = \sqrt{\dfrac{2F\cos\theta d}{m}}$.

This gives $v = \sqrt{\dfrac{2(35\text{ N})\cos 25°(6.5\text{ m})}{62\text{kg}}} = 2.6\frac{\text{m}}{\text{s}}$.

Note the shorter route to a solution using the work-energy theorem:

Focus: $v = ?$

Comment: Work done on the sled changes its KE. Only the horizontal component of the force contributes to the work.

From $W = \Delta\text{KE} \Rightarrow (F\cos\theta)d = \frac{1}{2}mv^2 \Rightarrow v = \sqrt{\dfrac{2F\cos\theta d}{m}}$

This gives $v = \sqrt{\dfrac{2(35\text{ N})\cos 25°(6.5\text{ m})}{62\text{ kg}}} = 2.6\frac{\text{m}}{\text{s}}$.

Can you see why physicists prefer the work-energy theorem in solving problems that involve force and distance?

Energy Problems

7-1. Tsing does work W in moving a box of mass m over a distance x on horizontal surface.
   (a) What force does Tsing exert in pushing the box?
   (b) Calculate the force that Tsing exerts when doing 660 J of work to move a 28-kg box over a horizontal distance of 11 m.
   (c) Calculate the net force on the box if friction between the box and the floor is a steady 60 N.

7-2. A vertical average force F applied over a height h is required to accelerate an elevator of mass m from rest up to a steady speed v.
   (a) How much work is done on the elevator by the applied force?
   (b) What is the kinetic energy of the elevator at steady speed v?
   (c) What is the potential energy of the elevator at height h?
   (d) What average force acts on an 850-kg elevator that is accelerated from rest to a speed of 2.0 m/s over a distance of 2.4 m?

7-3. Milo pushes a lawn mower of mass m, exerting a force F on the handle that makes an angle θ with the horizontal.
   (a) How much work does Milo do on the mower when pushing it a horizontal distance x?
   (b) Calculate the work Milo does on the mower if he pushes the 25-kg mower a distance of 15 m with a force of 43 N while the angle the mower handle makes with the ground is 45°.

7-4. A truck of mass m moves at speed v when moving against a headwind of speed V, while the temperature of the road surface is T.
   (a) What is the KE of the truck?
   (b) Calculate the kinetic energy of a 3300-kg truck moving at 18 m/s when a wind of speed 5 m/s hits it head on, and when the temperature of the road is 24°C. (Hint: Do you need all the information provided here?)

7-5. A battleship of mass M travels a distance x at constant speed in time t.
   (a) What is the KE of the battleship?
   (b) Calculate the KE of a 9.0×10^7-kg battleship that travels 250 km in 4.0 hours at a constant speed.

7-6. Pablo bats a baseball that weighs w and leaves the bat with a speed v.
   (a) What is the baseball's KE?
   (b) Calculate the KE of a 1.5-N baseball when it leaves the bat with a speed of 38 m/s.

7-7. A particle of mass m is accelerated from rest to a speed v by a steady force F. Assume that friction is insignificant in this case.
   (a) Over what distance does the force act?
   (b) Calculate the distance over which a 14-N force must act to change the speed of a 0.60-kg particle from rest to 12 m/s.

7-8. When an average force F is exerted over a certain distance on a system of mass m, its kinetic energy increases by $\frac{1}{2}mv^2$.

(a) Find the distance over which the force acts.

(b) If twice the force is exerted over twice the distance, how does the resulting increase in kinetic energy compare with the original increase in kinetic energy?

7-9. A constant horizontal net force F acts through a horizontal distance x on a cart of mass m, initially at rest.

(a) What is final kinetic energy of the cart?

(b) What is its final speed?

(c) Calculate answers to the above questions if the constant net force is 75 N, the mass of the cart is 8.4 kg, and the force is applied through a distance of 0.80 m.

7-10. A car of mass m undergoes an increase in speed from v_0 to v in a time interval t.

(a) What is the increase in its kinetic energy?

(b) Calculate the change of kinetic energy if the mass of the car is 1300 kg, its initial speed is 13 m/s, and 5 seconds later its speed is 19 m/s.

7-11. A bullet of mass m traveling at speed v hits a tree and slows down uniformly to a stop while penetrating a distance x into the tree trunk.

(a) What average force was exerted on the bullet in bringing it to rest?

(b) Calculate the average stopping force for a 5.2-gram bullet that hits the tree at a speed of 350 m/s and penetrates 12 cm into the trunk.

7-12. Ellyn throws a dart at a speed v into a thick dartboard. It penetrates a distance x.

(a) If she wants the dart to penetrate twice as far, how fast does it have to be thrown?

(b) How far would the dart penetrate if thrown at twice its initial speed? Why are your answers different?

(c) A dart thrown at 3.2 m/s penetrates 0.90 cm into the dartboard. Calculate the speed of a dart that would penetrate 1.8 cm into the dartboard.

(d) Calculate how far the same dart would penetrate if it were thrown at 6.4 m/s.

7-13. An automobile traveling on a level road at speed v brakes to a stop in a distance x.

(a) Assuming the same braking force, what would be the stopping distance for an initial speed of $4v$?

(b) Calculate the new stopping distance if the braking force brings a car with an initial speed of 20 km/h to rest in a distance of 12 m.

(c) From the above information, can the mass of the automobile be found?

7-14. A high diver of mass m dives from height h to the water below.

(a) What is her kinetic energy when she has freely fallen halfway to the water?

(b) What is her kinetic energy just before hitting the water?

(c) Calculate these values for a diver of mass 55.0 kg and initial height of 20.0 m above the water.

7-15. A window washer suspended from a scaffold at the top of a 35-floor skyscraper accidentally drops her scrub brush of mass m. The vertical distance between floors is y (so she is at height $35y$ when she drops the brush).

(a) What is the gravitational PE (relative to the ground below) and KE of the brush when it reaches the top of the 32nd floor? (Neglect air resistance.)

(b) If each story is 3.2 m, calculate the KE and PE of the 0.50-kg brush relative to the ground when it reaches the top of the 32nd floor.

7-16. A bucket of cement of mass m is lifted straight up by a crane to the third floor of a building under construction. The bucket is raised at a slow constant velocity v to an elevation h in the absence of air resistance and friction.

(a) How much work is done in raising the bucket? (Neglect its initial short acceleration.)

(b) The cable breaks just as the bucket reaches elevation h. What is the PE and the KE of the bucket at that point?

(c) Assuming no friction or air resistance as it falls, what is the total energy of the bucket when it has fallen halfway to the ground?

(d) What is the total energy of the bucket just before it hits the ground?

(e) What becomes of its energy upon hitting the ground?

(f) Calculate answers to parts (b), (c), and (d) for a 650-kg bucket that is moving at 0.5 m/s when it reaches a height of 15 meters.

7-17. Paul Doherty is rock climbing in the Grand Teton mountains and has just climbed to a height h when he drops his hammer of mass m. (Neglect air resistance.)

(a) Assuming Paul and the hammer were at rest at the time of the drop, what is the gravitational PE and the KE just when the hammer was dropped?

(b) What are the PE and the KE of the hammer after it has dropped 3/4 the way to the ground?

(c) What is the velocity of the hammer after it has dropped 3/4 the way to the ground?

(d) What is the gravitational PE and the KE of the hammer just before it hits the ground?

(e) What is the velocity of the hammer just before it hits the ground?

(f) What happens to the energy of the hammer after it hits the ground?

(g) Calculate answers to parts (b) through (e) for a 0.86 kg hammer dropped from a height of 44 m.

7-18. Marilyn the diver of mass m steps off a tower of height y above the water. After hitting the water she comes to a stop at distance d beneath the surface.

(a) What average net force does she experience while stopping beneath the water?

(b) Calculate the force if her mass is 50 kg, the tower 10 m above the water, and the distance to stop beneath the water is 2.0 m.

7-19. Paul's box of wrenches of mass m is located on a shelf of height y above the floor.

(a) What is the gravitational potential energy of the box relative to the floor when it is on a shelf of height $1.5y$?

(b) Calculate the potential energy of a 2.5-kg box of wrenches on the higher shelf if you know that the wrenches are made of high-quality steel, and $y = 1.2$ m.

60

7-20. A block of ice, mass m, at an initial height h slides from rest down an inclined plane. At the bottom of the incline it slides onto the floor at speed v.
   (a) Find this speed, assuming friction can be ignored.
   (b) Find the speed that a 27-kg block of ice will have when it reaches the bottom of an inclined plane of height 1.5 m. Again, ignore friction.

7-21. A chunk of ice of mass m breaks loose from a suspension bridge of height h.
   (a) What will be the chunk's kinetic energy when it hits the water below?
   (b) What would be its kinetic energy if it fell from a bridge twice as high?
   (c) A 0.80-kg chunk of ice falls from a suspension bridge 45 m above the water. Calculate the KE of the ice right before it hits the water.

7-22. Diane tosses a baseball straight upward with a certain speed, and therefore, a certain kinetic energy.
   (a) If air drag is negligible, what will be the ball's kinetic energy when caught at the same elevation from which it was thrown?
   (b) If air drag is not negligible, how will the kinetic energy differ when being caught?
   (c) Could you calculate the mass of the ball in part (a) if you're given its initial speed and the height to which it goes? Defend your answer.

7-23. A ball of mass m is thrown vertically upward with an initial speed v from a starting point y_0 above the ground. (Ignore air resistance.)
   (a) What is the maximum height reached by the ball?
   (b) What is the potential energy of the ball at its maximum height relative to the ground?
   (c) What is its potential energy relative to its launch point?
   (d) Calculate answers to the above questions for a 0.15-kg ball thrown upward with an initial speed of 23.0 m/s from a starting point 1.2 m above the ground.

7-24. A pendulum bob of mass m is pushed along its arc until it is a vertical distance y_1 above a table. After it is released, it swings to a lowest point y_2 above the table, then almost to its initial height y_1 on the other side.
   (a) By how much does the bob's PE at its highest point exceed its PE at its lowest point?
   (b) What is the kinetic energy of the bob as it passes through its lowest point?
   (c) Calculate answers to the above questions if the mass of the bob is 1.0 kg, the highest point is 72 cm above the table, and the lowest point 31 cm above the table.
   (d) If the angle that the pendulum string makes with the vertical is 45° at the pendulum's highest point, calculate the length of the string.

7-25. Alex drops a steel ball of mass m to the floor from a height h_1. The ball rebounds to a lesser height h_2.
   (a) How much kinetic energy does the ball have when it first bounces up from the floor?
   (b) How much potential energy did the ball have initially?
   (c) In your answers to parts (a) and (b), how do you account for the differences in energy?
   (d) A 0.50-kg ball falls from a height 1.2 m, hits the floor, and bounces to a height 0.90 m. Calculate the kinetic energy of the ball immediately after it bounces from the floor.

7-26. Gracie stands on a bridge of height h above a river and drops a stone of mass m.
(a) What is the stone's potential energy at the time of release relative to the river below?
(b) Neglecting air resistance, with what speed does the stone hit the water?
(c) What are the stone's kinetic and potential energies after it has fallen $h/2$?
(d) Calculate your answers to the above questions for a stone of mass 1.0 kg and a bridge height of 120 m.

7-27. A railroad car of mass m and speed v collides with a stationary car of equal mass. The cars couple and move together.
(a) What is the speed of the coupled cars after the collision? (Hint: Use momentum conservation.)
(b) What is the kinetic energy of the coupled cars?
(c) Calculate the kinetic energy of the coupled cars if each of them has a mass 10^4 kg and the initial speed of the moving car before collision was 5 m/s.
(d) How does the kinetic energy of the coupled cars compare with the initial kinetic energy of the first car?
(e) What explanation can you offer if there is an energy difference?

7-28. An acrobat of mass m stands on the left end of a seesaw. A second acrobat of mass M jumps from a height h onto the right end of the seesaw, thus propelling the first acrobat into the air.
(a) Neglecting inefficiencies, how will the PE of the smaller acrobat at the top of his trajectory compare with the PE of the person just as he jumps?
(b) Neglecting inefficiencies, how high does the first acrobat go?
(c) Calculate the maximum height the first acrobat reaches if his mass is 40 kg, the mass of the second acrobat is 70 kg, and the initial height jumped from was 4 m.

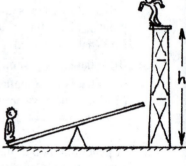

7-29. A nice demonstration utilizes a test tube of mass M suspended by a pair of parallel wires. The tube is sealed by a stopper of mass m. Inside the tube is a bit of water that is heated with a flame. When the water turns to steam the cork pops out, and the test tube swings back. The test tube rises a vertical distance h.
(a) What must have been the speed of the test tube immediately after the cork popped out?
(b) With what speed did the cork fly out?
(c) Why is momentum conservation part of the solution?

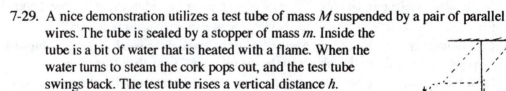

7-30. A pendulum bob of mass m is pulled back by an angle θ and then released.
(a) What is the speed of the bob when it reaches to the lowest part of its swing?
(b) At the bottom of its swing the pendulum bob has an elastic collision with a block of the same mass. Immediately after the collision, what is the speed of the pendulum bob?
(c) Neglecting friction, what is the resulting speed of the block?

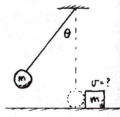

7-31. A roller coaster is pulled to the top of a hill of height h and allowed to roll down from rest. For a bigger thrill you might wish the car to be going twice as fast at the bottom of the run.
(a) How high does the hill need to be?
(b) How fast would the coaster go at the bottom if the hill were twice as high?
(c) Calculate the speed of a roller coaster at the bottom of a hill after starting from rest at the top of a 42-m hill.

7-32. A roller coaster of mass m starts from rest at an elevation h_1 and rolls down and then up another incline of lower height h_2.
(a) What is the kinetic energy of the coaster when it reaches the top of the second incline?
(b) Calculate the kinetic energy of a 2500-kg coaster atop the second incline of height 20.0 m if it starts from rest at the top of the first incline, 30.0 m high.

7-33. A roller coaster car of mass m is pulled from rest up to the top of a hill of height h where it is released with an initial speed very close to zero.
(a) What is the speed of the car when it rolls down to a distance $h/2$ above the ground?
(b) What force F will be necessary to stop the car in a distance d once it reaches the bottom of the hill?
(c) What would be the speed $h/2$ above the ground if the car had an initial speed v_0 at the top? (Hint: You can't just add v_0 to your earlier answer!)

7-34. Little Andre of mass m climbs a vertical ladder of height h to the top of a 45°-inclined plane of length L. The friction force acting on Andre as he slides down the plane is f.
(a) What is Andre's speed when he reaches the bottom of the inclined plane?
(b) Calculate Andre's speed at the bottom of the incline if the height h is 2.0 m and the coefficient of sliding friction μ_k is 0.30.
(c) If Andre's mass were 20% more than it is, how would this affect the speed?

7-35. A loaded elevator is lifted a distance of h in a time t by a motor delivering power P.
(a) How much force does the motor exert to lift the elevator?
(b) Calculate the force exerted if the distance raised is 20 m, the time of lift is 30 s, and the power of the motor is 60 kW.

7-36. A diesel engine lifts a pile-driver hammer of mass m to its maximum height h in a time t.
(a) What is the power output of the engine?
(b) Calculate the power delivered by the engine if the mass of the pile-driver hammer is 190 kg, and the height raised is 25 m in a time of 6.0 s.

7-37. An electric motor lifts an elevator of mass m from ground level to the top of a building of height h in time t.
(a) What average power does the motor deliver?
(b) Calculate the power the motor delivers when it lifts a 10,000-kg elevator from ground level to the top of a 30-m building in 40 s.

7-38. Equipment having mass m is raised a vertical distance h at a constant speed v to the top of the world's tallest lighthouse.
(a) How much power is required to raise the equipment?
(b) Calculate the power developed if the mass of the equipment is 25 kg, the vertical speed 2.0 m/s, and the height of the lighthouse is 106 m.

7-39. A motor of power P raises an elevator of mass m.
(a) What is the maximum speed at which the motor can raise the elevator?
(b) The elevator has a mass 900 kg and is powered by a 100-kW motor. Calculate the maximum speed at which the elevator can be raised.

7-40. Howie finds that a force F is required to push a crate of mass m up a plank of length L into a truck whose platform is a vertical distance h above the road.
(a) How much work does Howie do in pushing the crate up the plank?
(b) Calculate Howie's work input if the crate has a mass of 100 kg, the length of the plank is 5.0 m, the applied force is 490 N, and the platform is 1.2 m above the road.
(c) What is the increase of potential energy of the crate once on the platform?
(d) How much work did Howie do in overcoming friction?
(e) What is the efficiency of the plank?

7-41. Benjamin applies a force F over a distance x to a rope connected to a set of pulleys. This raises a block of weight W to height h above its initial level.
(a) What is Benjamin's work input?
(b) What work is done on the block?
(c) What is the efficiency of the pulley system?
(d) Calculate answers to the above questions given that the force applied is 75 N over a 12-m distance, the weight of the block is 300 N, and is raised a vertical distance of 2 m.

7-42. The block in the previous problem is raised a vertical distance h by means of a device having an efficiency E.
(a) What is the work output?
(b) What is the work input?
(c) Calculate answers to the above questions when the height raised is 5 m and the efficiency of the device is 65%.

Springs

The treatment of springs in the textbook is minimal, with Hooke's law first appearing in Chapter 12. Since stretching or compressing a spring requires work, we introduce a brief treatment of springs here.

Suppose you're in lab, measuring the stretch and force of a vertical spring suspended from a support. You place a ruler behind the spring, then hang a 50-gram mass at the end of the spring. Assume that you measure the spring to stretch 2 cm. Then you hang an additional 50 grams on the spring, and you find that it stretches another 2 cm.

If you were to enter your lab data in a table, it might (ideally) look like this:

Total mass hanging on the spring (grams)	Total stretch of the spring (cm)	Total mass hanging on the spring (kg) $\left[g \times \left(\frac{1kg}{1000g}\right)\right]$	Force pulling on the spring (N) $m_{hanging}g \quad \left(\text{Use } g = 10\frac{m}{s^2}\right)$	Total Stretch of the spring (m) $\left[cm \times \left(\frac{1m}{100cm}\right)\right]$
0	0	0	0	0
50	2	0.050	0.5	0.02
100	4	0.10	1.0	0.04
150	6	0.15	1.5	0.06
200	8	0.20	2.0	0.08
250	10	0.25	2.5	0.10

(Notice that the numbers are easier to "grasp" in grams and centimeters, which is why we often use these units in the lab.)

If you graph the last two columns of the data you can see that you'll get a straight line.[1] The straight line tells you that the two quantities, applied force and stretch, are directly proportional. Double one, the other doubles, and so forth. The equation of a straight line through the origin is

$$y = slope \cdot x$$

The proportionality of force to stretch (or compression) was discovered by English physicist Robert Hooke, a contemporary of Isaac Newton. This is Hooke's law:

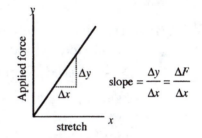

Applied force $F = kx$

The constant k is the *spring constant* (typically in N/m), which tells us how much force corresponds to how much stretch. Many springs and other "springy" objects exhibit a stretch that is proportional to the applied force, at least for small deformations.

Springs in physics problems have three properties:

1. They are *linear*—the amount of stretch is proportional to the amount of applied force.

2. Their behavior is *symmetric*—a given amount of force will compress or stretch a spring the same amount.

3. They are light compared with the objects we suspend or place on them, so we can *approximate* them as having zero mass.

---

[1] It is common practice to place the experimental variable we control—the *independent variable* (in this case the hanging weights)—on the x-axis, and the experimental variable that we measure—the *dependent variable* (the stretch of the spring)—on the y-axis. Here we reverse the practice.

© Paul G. Hewitt and Phillip R. Wolf   65

Sample Problem 1
When you stretch (or compress) a spring, it has potential energy, the ability to do work.
(a) How much work is required to stretch a particular spring a distance x by a force F?

Focus: $W = ?$

At first thought the solution might be, $W = Fd = Fx$.

Not so, for the full value of the force F doesn't occur until the very end of the stretch. Ordinarily, when you stretch a spring by a distance x, the initial force is zero and builds linearly to kx. So the average force exerted is $kx/2$. That means the work done in stretching the spring is

$$W = F_{average}x = \left(\tfrac{1}{2}kx\right)x = \tfrac{1}{2}kx^2.$$

The work done in stretching or compressing a spring is converted to *elastic potential energy*. This stored energy can be transformed to some other form.

(b) Calculate the work done in stretching a spring when a force of 50 N stretches it by 20 cm (0.20 m).

Solution: The solution is *not* $W = Fd = (50 \text{ N})(0.20 \text{ m}) = 10$ J. Rather, the work done is the *average* force × the distance (where the average force is one-half 50 N);

$$W = \tfrac{1}{2}(50 \text{ N})(0.20 \text{ m}) = \textbf{5 J, or using } k = \frac{50 \text{ N}}{0.20 \text{ m}} = 250 \tfrac{\text{N}}{\text{m}},$$

$$W = \tfrac{1}{2}kx^2 = \tfrac{1}{2}\left(250 \tfrac{\text{N}}{\text{m}}\right)(0.20 \text{ m})^2 = \textbf{5 J}.$$

Notice we get the solution two ways. It's nice that there are more ways than one to solve many physics problems!

Sample Problem 2
A mass m is compressed against a spring with a spring constant k on a frictionless surface.
(a) What is the speed v of the mass when it is released by the spring?

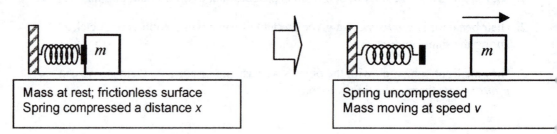

| Mass at rest; frictionless surface
Spring compressed a distance x | Spring uncompressed
Mass moving at speed v |

Solution: The work required to compress the spring originally was $\tfrac{1}{2}kx^2$. This elastic potential energy is transformed into KE of the mass.

$$\left(\text{Elastic PE in the Spring}\right)_{initial} = \left(\text{KE of the mass}\right)_{final}$$

$$\tfrac{1}{2}kx^2 = \tfrac{1}{2}mv^2 \implies v^2 = \frac{k}{m}x^2 \implies v = \sqrt{\frac{k}{m}}x.$$

Springs and Elastic-Potential-Energy Problems

7-43. In lab you have a box of 1-kg cylinders that can be hooked on to one another. You suspend one cylinder from a spring and record the spring's stretch. You continue suspending additional cylinders, one at a time, keeping measurements of the stretch.
(a) How does the stretch change per added cylinder?
(b) Sketch what a graph of added weight vs. stretch would look like for the spring.
(c) What does the slope of the graph indicate about the spring?

7-44. In lab you find the spring constant k of a certain spring by applying a force F and measuring the resulting stretch, a distance x.
(a) What is the spring constant?
(b) Calculate the spring constant if the stretching force is 5.0 N and the stretch of the spring is 6.5 cm.

7-45. A force F compresses a spring that has a spring constant k.
(a) How much is the resulting spring compression?
(b) Calculate the compression distance if a 10-N force is used to compress a spring that has a spring constant 200 N/m.

7-46. A spring of spring constant k is stretched a distance x from its equilibrium position.
(a) How much work is required to stretch the spring this distance?
(b) Calculate the amount of work required if the spring constant is 45 N/m and the distance of stretch from the equilibrium position is 2.50 cm.

7-47. Timbo does a certain amount of work W to stretch a spring a distance x.
(a) What is the spring constant?
(b) Calculate the spring constant if it takes 400 J of work to stretch a spring 8.0 cm from its equilibrium position.

7-48. • Timbo uses a spring to pull horizontally on a block of mass m that rests on a lab table. The spring constant k and the values of μ_k and μ_s are known quantities.
(a) How much will the spring have to be stretched to get the block to start moving?
(b) How much stretch is needed to keep the block moving at constant speed?
(c) If the spring is stretched by an amount x, more than the amount in part (b), find the acceleration of the moving block.
(d) Evaluate each of the above for $k = 55$ N/m, $x = 20.0$ cm, $m = 3.2$ kg, $\mu_s = 0.50$, and $\mu_k = 0.30$.
(e) Calculate the work done in stretching the spring a distance 20.0 cm.

7-49. Bungee jumper Margaret of mass m jumps from a bridge of height h above the ground. She is attached to a bungee cord of length L (less than h!) and spring constant k. One end of the cord is attached to the bridge and the other end to her.
(a) What is the minimum value that k must have so that Margaret just avoids hitting the ground?
(b) Calculate the minimum value of k for a 70.0-kg woman and a 32.0-m high bridge, with a 10.0-m long bungee cord.
(c) Calculate Margaret's maximum acceleration.

7-50. • A carnival game consists of a spring of spring constant k that you compress to fire a small block of mass m up a frictionless ramp at an angle θ and length L. A cup sits at ground level directly against the base of the far edge of the ramp.

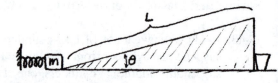

(a) How far should you compress the spring so that the block will fall into the cup?

(b) Calculate the amount of spring compression if $k = 10$ N/m, $m = 10$ grams, $L = 1.22$ m, and $\theta = 15°$.

7-51. • Place a block of mass m on top of a vertical spring of spring constant k mounted on the floor. Then push the block down a distance x from the spring's initial position.

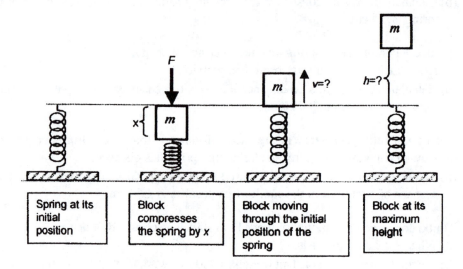

| Spring at its initial position | Block compresses the spring by x | Block moving through the initial position of the spring | Block at its maximum height |

(a) What force must you exert to hold the block at rest on the compressed spring?

(b) When you release the block, what is its speed when the spring has returned to its initial position?

(c) How high above the initial position of the spring will the block rise?

7-52. • A force F pulls at an angle θ above the horizontal on a block of mass m. The values of μ_k and μ_s are known.

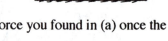

(a) Find the minimum value of F that will get the block to start moving.

(b) Find the acceleration of the block if we maintain that force you found in (a) once the block starts moving.

(c) How much work is done in moving the block a horizontal distance x?

Show-That Problems for Energy

7-53. Marshall applies a constant force of 20-N to a box and causes it to move at a constant speed of 4.0 m/s.
Show that the work he does in moving the box in 6.0 s is 480 J.

7-54. A 0.50-kg sphere at the top of an incline has a potential energy of 6.0 J relative to the base of the incline.
Show that when the sphere rolls halfway down the incline its potential energy is 3.0 J.

7-55. A 20-N block falls freely from rest from a point 3.0 m above the surface of Earth.
Show that the gravitational potential energy of the block after falling 1.5 m is 30 J.

7-56. A cart moving at a constant speed of 25 m/s possesses 438 J of kinetic energy.
Show that the cart has a mass of 1.4 kg.

7-57. A steady force of friction acts on a 15.0-kg mass moving 10.0 m/s on a horizontal surface.
Show that if the mass is brought to rest over a distance of 12.5 m, the friction force is 60.0 N.

7-58. Mike's vintage sports car travels along a level road at 64 km/h, and skids to a stop with pre-antilock brakes.
Show that if its speed were 96 km/h, its skidding distance would be 2.25 times as much.

7-59. Mark pulls a 55-gram arrow back 71 cm on a bow that exerts an average force of 100 N on the arrow.
Show that the release speed is about 51 m/s.

7-60. A 1-kg projectile is fired at 10 m/s from a 10-kg launcher. The momenta of the projectile and the launcher have the same magnitude, but opposite directions.
Show that the KE of the projectile is ten times greater than the KE of the recoiling launcher.

7-61. A railroad boxcar traveling at speed 4 m/s has an inelastic collision with an identical boxcar at rest. The coupled cars then move together at 2 m/s.
Show that the amount of energy converted to thermal energy in the collision is half the initial KE of the moving car.

7-62. Dan finds that 700 watts of power is needed to keep his boat moving through the water at a constant speed of 10 m/s.
Show that the magnitude of the force exerted by the water on the boat is 70 N.

7-63. Sue launches an 8.0-kg boulder from a makeshift slingshot. The rubber bands making up the slingshot are stretched 2.2 m and the boulder is launched at a near-horizontal speed of 12.0 m/s.
Show that the spring constant of the combined rubber bands is 240 N/m.

7-64. • A golf ball is hurled at and bounces from a very massive bowling ball that is initially at rest.
Show that after the collision the golf ball has about the same kinetic energy, and that the bowling ball has nearly twice the initial momentum of the golf ball.

8 Rotational Motion

In this chapter we distinguish between linear and rotational speeds. Linear speed v, as covered in Chapter 3, is motion measured in units of meters/second. When motion follows a curved path we speak of tangential speed, the speed that is tangent or parallel to the curve at a particular point. It also is measured in meters/second. The curved paths that we consider in this chapter are either circles or arcs of circles.

Rotational speed ω is measured in RPM, rotations per minute. Or it can be rotations per second, rotations per month, or some number of complete rotations in a given time. Or it can be some angle per unit time.

Imagine that you are sitting on a rotating turntable. The faster it spins, the greater distance you travel each second. Put another way, the greater its rotational speed, the greater your tangential speed:

$$v \sim \omega.$$

Sitting farther away from the axis of the spinning turntable is another way to cover more distance in the same amount of time:

$$v \sim r.$$

Combining both of these ideas, we say:

$$v \sim r\omega.$$

We can use the exact equation $v = r\omega$ only if we use SI units for measuring ω. The SI unit for ω is radians/second, where 1 radian is an angle approximately equal to 57°. In keeping with the spirit of the textbook chapter we've chosen not to include problems where using radians is necessary for a solution.

We say that an object following a circular path is accelerating because its velocity is changing continually—perhaps not in magnitude, but certainly in direction. We call the acceleration toward the center of a circle *centripetal* acceleration. It is defined by:

$$a_{centripetal} = \frac{v^2}{r}, \text{ or equivalently,}$$

$a_{centripetal} = r\omega^2$ (again in radians!). In this book we go with $\frac{v^2}{r}$.

The force producing the circular motion in any given situation is called the *centripetal force* and has magnitude $ma_{centripetal}$, or more specifically, $F_c = m\frac{v^2}{r}$.

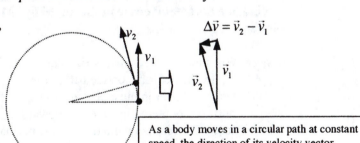

$$\Delta \vec{v} = \vec{v}_2 - \vec{v}_1$$

As a body moves in a circular path at constant speed, the direction of its velocity vector changes. From $\vec{a} = \frac{\Delta \vec{v}}{\Delta t} = \frac{\vec{v}_2 - \vec{v}_1}{\Delta t}$, the acceleration vector points in the same direction as Δv, toward the center of the circle.

Whenever something moves in a circular path there must be some force pushing or pulling it towards the center of the circle—otherwise it would fly off along a straight-line tangential path. In a problem involving rotational motion, first identify the physical force that pushes or pulls the rotating object towards the center, then set it equal in magnitude to mv^2/r and solve. Sometimes there is more than one force in the radial direction, in which case you set the *net* force to mv^2/r. Voila!

Sample Problem 1

Suppose you attach a ball of mass *m* to a string and whirl it overhead in a horizontal circular path. If you keep the ball moving with constant linear speed *v* and shorten the string, tension in the string increases. The string will break if you exceed a critical tension *T*.

(a) What is the shortest length of string you can use so the string doesn't break? (Make the approximation that the string remains horizontal.)

Solution: Step 1. *Focus*: Length of string $L =$?

Step 2. The physics concept here is centripetal force, $F = \dfrac{mv^2}{r}$. String tension T

provides the centripetal force, so $F = T = \dfrac{mv^2}{L}$.

Step 3. From $T = \dfrac{mv^2}{L} \Rightarrow L = \dfrac{mv^2}{T}$.

If we plug in the *maximum* value T can have, this gives us the *minimum* value L can have for a given mass and speed.

(b) Calculate the shortest length of the string you could use to keep a 250-gram ball circling at a constant speed of 6.0 m/s if the string's breaking strength is 45 N.

Solution: First convert to consistent SI units: $250 \, \text{g} \times \frac{1 \, \text{kg}}{1000 \, \text{g}} = 0.25 \, \text{kg}$.

Then $L = \dfrac{mv^2}{T} = \dfrac{(0.25 \, \text{kg})(6.0 \, \frac{\text{m}}{\text{s}})^2}{45 \, \text{N}} = 0.20 \dfrac{\text{kg} \cdot \frac{\text{m}^2}{\text{s}^2}}{\text{kg} \cdot \frac{\text{m}}{\text{s}}} = \textbf{0.20 m}$.

(c) The string could remain horizontal if the ball were on a horizontal surface, but not if it's moving in a horizontal circle in the air. Why? Make a rough sketch to support your answer.

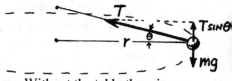

Answer: When a ball moves in a horizontal circle, the net force on the ball in the vertical direction must be zero, so *mg* must be countered by an equal upward force. On a horizontal table it is the normal force. Without the table there is no normal force and the upward force must be supplied by a component of the string tension. Therefore the string must make an angle with the horizontal as shown in the diagram.

Sample Problem 2
A flyer can experience weightlessness by doing a
vertical loop-the-loop at just the right speed.
(a) For a vertical loop of radius r, at what critical
speed will the pilot feel weightless?

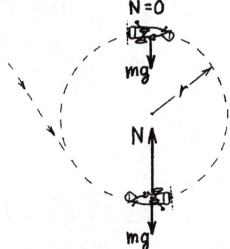

   Solution: Step 1. *Focus:* $v = ?$

           Step 2. The pilot is in circular motion, so
              this is a centripetal-force problem.

$$\text{From } F = \frac{mv^2}{r} \implies v = \sqrt{\frac{Fr}{m}}.$$

Two principal forces act on a flyer while executing a
vertical loop, gravitational force mg and the normal force
N from the seat pushing on the pilot's body. At the top of the loop both forces are in the radial
direction (downward, in this case). Together they provide the centripetal force.

The net force $F_{net} = mg + N = \dfrac{mv^2}{r}$.

When the pilot experiences weightlessness, the normal force is zero ($N = 0$), so the net force
$F_{net} = mg$.

$$\text{Step 3. } v = \sqrt{\frac{Fr}{m}} = \sqrt{\frac{mgr}{m}} = \sqrt{gr}.$$

(b) Calculate the critical speed for achieving weightlessness at the top of a vertical loop of
radius 200.0 m.

   Solution: $v = \sqrt{gr} = \sqrt{(9.80 \text{ m/s}^2)(200.0 \text{ m})} = \textbf{44.3 m/s}.$

(c) If the plane makes the entire vertical loop at a constant speed, where along the loop will
the pilot feel most heavy? Defend your answer with a sketch showing relative
magnitudes of N and mg.

   Answer: Normal force N is a measure of how heavy one feels. The pilot feels heaviest at
        the bottom of the loop because N there is a maximum. Whereas mg alone
        provides the centripetal force at the top when N is zero, the centripetal force at
        the bottom of the loop is provided by $N -$
        mg. The sketch indicates that N at the
        bottom must be $2mg$ if N at the top is zero
        and speed is the same in both locations. If
        the speed is greater at the bottom, N will
        be even larger. (At all points along the
        loop only mg and N act on the pilot. At
        locations other than the top and bottom,
        the centripetal force is provided by the
        components of N and mg that lie along the
        radial direction—another story. Would a
        parabolic curve provide a longer period of
        weightlessness? That's still another story.)

Sample Problem 3

A space pod at the end of a tether line of length L moves at speed v_0 in a circular path about a space station.

(a) What will be the speed of the pod if the length of the line is reduced to 0.25 L?

Focus: $v = ?$

The only force acting on the pod is the tension in the tether. The tension force is directed radially inward, toward the center of the pod's circular path. Hence the tension does not exert a torque on the pod, which means the angular momentum of the pod is conserved.

$$(\text{Angular momentum})_{\text{initial}} = (\text{Angular momentum})_{\text{final}}$$

$$mv_0 L = mv_{\text{new}} r_{\text{new}} = mv_{\text{new}}(0.25L)$$

Solution : $v_{\text{new}} = \dfrac{v_0}{0.25} = 4v_0$.

(b) If the initial speed of the space pod is 1.0 m/s, what is the speed when pulled in one-fourth of its original distance from the space station?

Solution: $v_{\text{new}} = 4\left(1.0\frac{m}{s}\right) = 4.0\frac{m}{s}$.

Torque

In Chapter 2 we learned the equilibrium rule: $\Sigma F = 0$. Now we complement the rule with the added condition $\Sigma \tau = 0$ (sum of the torques = 0). In its simplest form, $\tau =$ lever arm $\times$ force. When the force and the lever arm are not mutually perpendicular trig comes into the picture.

Consider the following three cases involving a heavy, hinged beam. The hinge is attached to the wall. In each case the weight of the bar exerts a torque tending to make the bar rotate clockwise.

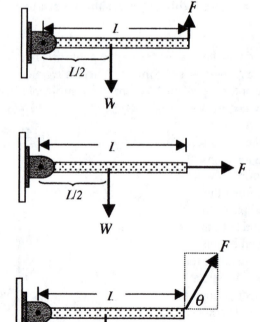

Case 1: *All* of the force F is exerted perpendicular to the lever arm. The torque due to F is FL.

Case 2: *None* of the force F is exerted perpendicular to the lever arm. The torque due to F is zero (in this case the beam would rotate clockwise due only to W).

Case 3: The *component* of F exerted perpendicular to the lever arm will cause a torque on the beam. This component has a magnitude $F\sin\theta$, so the torque due to F is $(F\sin\theta)L$.

Notice in case 3 that we take the component of F perpendicular to the lever arm. We would come to the same result if we instead multiplied F by the component of the lever arm that is perpendicular to the line of action of F (examples in the text illustrate this later method). Both methods are equivalent.

Sample Problem 4
A sign with mass m is centered on a lightweight bar of length L that is hinged at one end. A cable is attached to the other end of the bar and to the wall as shown.
(a) Find the tension in the cable?

Focus: Cable tension $T = ?$

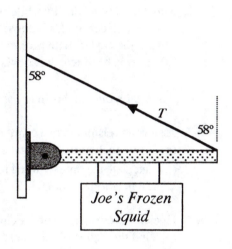

Since the bar is at rest the net torque acting on it must be zero. Clockwise torque about the hinge is provided by the weight of the sign, which we assume acts at the center of the bar (since the bar is lightweight, we assume that it doesn't contribute to the torque). Counter-clockwise torque is provided by the tension in the cable. The two torques added algebraically equal zero.

The clockwise torque due to the weight of the sign is simple enough—simply $W(L/2)$. The counterclockwise torque due to cable tension acts at a distance L from the hinge and is provided by the component of T acting perpendicular to the bar, $T\cos\theta$.

So $T\cos\theta\, L - W\left(\frac{L}{2}\right) = 0 \quad \Rightarrow \quad T = \dfrac{mg}{2\cos\theta}$.

(b) Suppose the mass of the sign is 1.3 kg and the bar is 1.2 m long. Calculate the tension in the cable.

Solution: Cable tension $T = \dfrac{(1.3\,\text{kg})\left(9.8\frac{\text{m}}{\text{s}^2}\right)}{2\cos 58°} = \textbf{12 N}.$

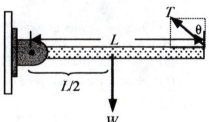

Circular Motion Problems

8-1. Little Megan rides on a horizontal rotating platform of radius r at an amusement park and feels the air passing through her hair as she moves at speed v. She rides at a location one-third the way from the center to the outer edge.

   (a) If the rotation rate of the platform remains constant, what will be her linear speed if she moves to the outer edge?

   (b) Calculate Megan's speed at the outer edge if she was moving at 1.0 m/s when she was at one-third the radius from the center.

8-2. Manuel rides at speed v while in the seat of a rotating Ferris wheel. Horsing around, Manuel decides to climb down a spoke of the wheel. He climbs one-quarter the distance toward the center, with people below astounded.

   (a) What is his linear speed when located three-quarters the distance from the rotational axis?

   (b) What would be his linear speed at the central axis?

8-3. Joshua travels in his racecar at a steady speed v around a circular track of radius r.

   (a) How much time does it take to make a complete circle?

   (b) Calculate the time required to go around a track of radius 134 m at a speed of 60 m/s.

8-4. Chelcie races once around a circular track at a speed v in t seconds.

   (a) Find the radius of the track.

   (b) Calculate the radius for a speed of 5.5 m/s and a time of 73 s to complete one circle.

8-5. Norman standing on Earth's equator has a tangential speed due to Earth's daily spin.

   (a) Given that the mass of Earth is M, and radius R, what is the tangential speed of Norman and his friends at the equator?

   (b) Calculate the tangential speed at Earth's equator. Earth's equatorial radius is 6.37×10^6 m.

8-6. A ceiling fan rotates at n rpm (revolutions per minute). The tips of the fan are distance r from the axis.

   (a) What is the linear speed of the tips of the ceiling fan?

   (b) Calculate the linear speed of the tips if they are 0.66 m long and turn at 15 rpm.

Rotational Inertia Problems

8-7. Consider three objects: Object A is a hoop of mass m and radius r; Object B is a sphere of mass $2m$ and radius r; Object C is a disk of mass $3m$ and radius r.

   (a) Which has the greatest rotational inertia?

   (b) Which has the smallest rotational inertia?

8-8. When a block initially at rest slides down an incline of negligible friction, the speed at the bottom is $\sqrt{2gh}$ (as found when PE at the top transfers to KE at the bottom). All blocks starting from rest, regardless of mass, will have equal accelerations on an inclined friction-free plane. Consider a race between a block on a friction-free incline, and a bowling ball on an incline that is identical, except that there is enough friction for rolling to occur. A ball rolling down an incline has "translational" KE, covered in Chapter 7, and *rotational KE*, a topic not covered in the textbook.

(a) Without knowing the details of rotational KE, which will win the race: the block or the bowling ball?

(b) Which would win the race if both were on friction-free same-angle planes (so that the ball would slide rather than roll)?

8-9. A playground carousel is free to rotate about its center on frictionless bearings. The carousel has a mass M, radius r, and is at rest. Consider the carousel to simply be a big disk.

(a) What is the rotational inertial of the carousel?

(b) What would be the rotational inertia of a man of mass m rotating in a circle of radius r (the same radius as the carousel)?

(c) It so happens that the rotational inertias of bodies add when they share the same center of rotation. What will be the combined rotational inertia of the man and carousel when the man sits at the edge of the carousel?

(d) Calculate the combined rotational inertia if the diameter of the carousel is 4.0 m, its rotational inertia alone is 400 kg·m², and the mass of the man is 70.0 kg.

Torque Problems

8-10. Mary Beth uses a torque feeler that consists of a meterstick held at the 0-cm end with forces applied to various positions along the stick. (See the photograph that opens Chapter 8 in your textbook.) In Trial A, force F pulls down at the 100-cm mark. In Trial B, force $2F$ pulls down at the 50-cm mark. In Trial C, $3F$ pulls down at the 40-cm mark. (Neglect the weight of the meterstick.)

(a) Which trial produces the greatest torque?

(b) Which trial produces the least torque?

(c) In which two trials are the torques equal?

8-11. James uses a wrench to turn a stubborn bolt.

(a) How much would the torque on the bolt increase if the applied force were doubled?

(b) How much would the torque increase for the original force if a pipe were inserted over the wrench to make the lever arm 3 times as long?

8-12. Jose pushes perpendicular to a door with force F at the door's far edge. Marie pushes the door from the opposite side and the door doesn't turn. Her push is perpendicular to the door but at a point halfway from the hinge to the edge. (A top view of the door is seen in the sketch.)

(a) How hard does she push on the door?

(b) Calculate Marie's push if Jose pushes with a force of 10 N.

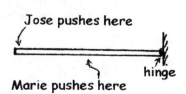

Jose pushes here

Marie pushes here

hinge

8-13. Marcus of mass M and Juanita of mass m balance each other on opposite sides of a seesaw. The fulcrum is located at the middle of the seesaw.

(a) If Marcus sits at distance D from the fulcrum how far does Juanita sit from the fulcrum?

(b) If Marcus weighs 360 N and sits 1.25 m from the fulcrum, where does 315-N Juanita sit?

8-14. A boy and girl sit on a seesaw of length L balanced at its center. The girl sits at the far end. The boy is twice as heavy as the girl, and therefore sits midway between his end and the center. Then the girl is given a bag of oranges weighing 10 N and the seesaw rotates out of balance. When the boy is given a bag of apples, balance is restored.

(a) What is the weight of the apples compared with the weight of the oranges?

(b) If the boy didn't have a bag of apples to increase his weight, how much farther from the center should he sit to restore balance?

8-15. Consider two identical metersticks tied together with a loose piece of string attached to their ends. If you hold one stick horizontal, the other dangles vertically, as shown. You place your finger at an appropriate location beneath the horizontal stick so the two sticks will balance.

(a) Where along the horizontal stick should you place your finger so the metersticks will balance?

(b) If you did the same with one horizontal meterstick and two vertical metersticks dangling from its one end, where would you place your finger to balance the system?

8-16. A uniform heavy plank of mass m lies on a raised horizontal surface, with one end overhanging. The maximum overhang distance is half the length of the plank. Suppose you put your pet cat of mass $0.2\,m$ on the end of the overhung plank (with a safety net below). Of course the plank would topple, unless it is moved to the left.

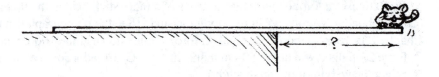

(a) What percentage of the plank can overhang without toppling, with the cat at its end?

(b) If the cat were twice as heavy, what would be the maximum percentage of overhang distance?

8-17. Lydia has a mass m and stands at the end of a uniform plank of length L and mass M that overhangs the top of a building as shown. The maximum distance of overhang for tipping to not occur is one-eighth the length of the plank.

(a) How does Lydia's mass compare with the mass of the plank?

(b) Calculate the mass of the plank if Lydia's mass is 45 kg.

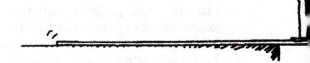

8-18. A ball rolls on an inclined plane because of an
unbalanced torque. Suppose the plane is inclined at
angle θ.

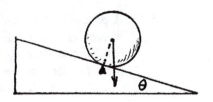

(a) On the drawing, sketch in the lever arm.

(b) Will an incline of 2θ provide twice as much torque
on the ball? Defend your answer.

Centripetal Force Problems

8-19. Consider three cases of whirling a puck in a circular path at the end of a string on a
friction-free air table. For Case A the linear speed of the puck is v and length of string L.
For Case B the linear speed of the puck is $0.5v$ and the length of string $2L$. For Case C
the linear speed of the puck is $1.5v$ and the length of the string is $1.5L$.
(a) In which case is tension in the string greatest?
(b) In which case is tension in the string least?
(c) Are there two cases where the tension is the same?

8-20. A toy electric train of mass m rounds a circular track of diameter D once in time t.
(a) How much centripetal force is provided by the tracks?
(b) Calculate the centripetal force acting on a 0.80-kg train rounding a 2.0-m diameter
track every 12 s.

8-21. A centripetal force is needed to keep a block of mass m moving on a smooth floor in a
circular path of radius r. The block makes a complete revolution in time t.
(a) Find the centripetal force.
(b) Calculate the centripetal force needed if the mass is 2.0 kg, the radial distance is 1.0 m,
and each revolution takes 5.0 s.

8-22. A model airplane of mass m flies in a horizontal circle of diameter D and completes
each round trip in time t.
(a) What is its speed?
(b) Calculate the speed if the plane has a mass of 0.80 kg, the diameter of the circular
path is 5.0 m, and the time to make a revolution is 2.0 s.
(c) Calculate the centripetal acceleration of the plane.
(d) Calculate the centripetal force that acts on the plane if its mass is 0.50 kg.

8-23. An air puck of mass m is attached to a string of length L and moves in a horizontal
circle.
(a) At what speed must the puck move so that the centripetal acceleration is equal to the
acceleration of gravity g?
(b) Calculate the puck's speed for a 0.60-m long string and a 0.20-kg puck.

8-24. A ball revolves in a horizontal circle at speed v at the end of
a string of length L. The tension in the string is T. (Assume
the string is approximately horizontal.)

(a) What is the mass of the ball?

(b) Calculate the mass when the string is 5.0 meters long, the
speed is 8.0 m/s, and the tension is 25 N.

8-25. A pellet-filled tin can is tied to the end of a cable of length L. The tension in the cable is T when the can revolves around a center pole in a horizontal circle at a steady speed v.
(a) Find the mass of the can.
(b) When the can moves at 15 m/s at the end of a 2.0-m long cable, the tension in the cable is 195 N. Calculate the mass of the can.

8-26. A centripetal force acts on an airplane of mass m flying at speed v in a horizontal circle of radius r.
(a) Find the centripetal force divided by the mass (this is the force per unit mass).
(b) Calculate the centripetal force per kilogram when the speed of the airplane is 200.0 m/s and the radius of its path is 12,500 m.

8-27. An electric force holds an electron of mass m in a circular orbit at a uniform speed v, and orbital radius r.
(a) What is the magnitude of electrical force holding the electron in orbit?
(b) Calculate the force, given that the mass of an electron is 9.1×10^{-31} kg, its orbital radius is 5.3×10^{-11} m, and it whirls about the nucleus at speed 2.2×10^6 m/s.

8-28. Marshall swings a pail of water in a vertical circle of radius r.
(a) What minimum speed must he give the pail at the top of its path so no water spills?
(b) Why does your answer not depend on the mass of the water?

8-29. A crate of mass m rests on the bed of a flatbed truck that moves at speed v around an unbanked curve of radius r.
(a) What coefficient of static friction between the crate and the flatbed is required to barely hold the crate from slipping off the flatbed?
(b) Calculate the coefficient of friction needed to hold the 100-kg crate when the speed of the truck is a constant 14.0 m/s and the radius of the curve is 60.0 m.

8-30. • Investigators find that a drop of water of mass m will barely adhere to a piece of cloth in a washing machine drum when it rotates at 1200 rpm.
(a) Find the force of adhesion between the drop and the cloth when circling at radius r.
(b) Calculate the force of adhesion if the mass of the drop is 5×10^{-5} kg and the radial distance of the cloth from the axis of the drum is 0.40 m.

8-31. • A cotton-fiber string attaches a ball of mass m to a rotating device. The string can exert sufficient force to keep the ball circling at a radius r when whirled by the device. The string snaps if the motion is increased.
(a) What is the minimum radius at which a ball of one-tenth mass m may be whirled by the same string at the same rotational speed in the same device without having the string break? (Assume that the string can be approximated as horizontal.)
(b) Calculate the radius if the initial radius was 2.10 m for a mass of 2.0 kg.

8-32. A model airplane flies in a horizontal circle at the end of a string that can at most supply tension T. The mass of the airplane is m and its speed is v. Assume that the speed is great enough so that the string is nearly parallel to the ground.
   (a) What is the smallest radius r at which the plane can be flown before the string breaks?
   (b) Calculate the radial distance if the plane's mass is 0.70 kg, its speed 30 m/s, and maximum tension is 175.0 N.

8-33. A penny sits on the outer edge of a phonograph record of diameter D. The penny moves with a speed v.
   (a) Find the minimum coefficient of static friction μ_s that will keep the penny from flying off the record.
   (b) Calculate μ_s if the speed of the penny is 0.60 m/s and the diameter of the record is 0.16 m.

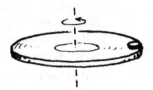

8-34. One of the earliest objections to the idea of a rotating Earth was the claim that people at the equator would be thrown off into space. For this to happen, "centrifugal force" would have to be greater than mg.
   (a) How much "centrifugal force" acts on a person at the equator? (Earth's equatorial radius is 6.37×10^6 m.)
   (b) Suppose you wish to calculate the "centrifugal" or centripetal force on a person at a latitude of 45° (halfway between the equator and the Earth's pole). In the centripetal force equation, would your radial distance be Earth's radius, or Earth's radius $\times \cos 45°$? Defend your answer.

8-35. At the equator (radius 6.37×10^6 m) you have a tangential speed due to Earth's daily rotation.
   (a) Suppose the spin rate increased. How large would your tangential speed have to be to fly off Earth?
   (b) Why would an Earth's increased spin not throw people off the Earth's poles?

8-36. A small battery-driven car of mass m moves at a constant speed v in a vertical circle on the track shown. The radius of the track is r.
   (a) What is the magnitude of the normal force acting on the cart at A, the highest point?
   (b) What is the magnitude of the normal force acting on the cart at C, the lowest point?
   (c) What is the magnitude of the normal force acting on the cart at B, the 3:00-o'clock position halfway up the track?

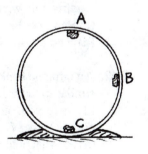

Angular Momentum Problems

8-37. A space pod circles a space station at the end of a long tether line.
   (a) How will the linear speed of the pod compare with its original speed when the line has been pulled in to half its length?
   (b) How will the linear speed compare when the line is pulled in to one-tenth its original length?

8-38. A small space telescope at the end of a tether line of length L moves at linear speed v about a central space station.
(a) What will be the linear speed of the telescope if the length of the line is reduced to $0.33\,L$?
(b) If the initial linear speed of the telescope is 1.0 m/s, what is its speed when pulled in to one-third its initial distance from the space station?

8-39. A ball of mass m at the end of a string moves at constant speed v in a circle of radius r. Assume the speed is great enough so the string remains practically horizontal.
(a) If the ball is pulled into a circular path of half the radius, what will be its speed?
(b) By how much will the tension have increased when the ball is at this new radius?
(c) Calculate the increased tension if the mass of the ball is 0.50 kg, initial speed 20 m/s, and initial radius 4.0 m.

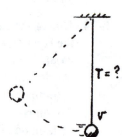

8-40. A bowling ball of mass m is attached to the ceiling by a rope of length L. The ball is pulled to the side and released, swinging to and fro. At the bottom of the swing the ball has speed v.
(a) What is the tension in the rope at the bottom of its swing?
(b) Calculate the rope tension at the bottom of the swing for a 36-kg ball, a 2.0 m long rope, and a speed at the bottom of 4.4 m/s.

Problems with a Bit of Trigonometry

8-41. A popular lab activity that illustrates a conical pendulum is the toy Flying Pig (featured in the Lab Manual). The sketch shows two forces acting on the Pig; tension T in a string of length L and weight mg.
(a) How does the centripetal force on the Pig compare with the horizontal component of T?
(b) Since the Pig isn't accelerating in the vertical direction, what is the vertical component of T?
(c) How does tension T compare with the weight of the Pig?
(d) Find the speed of the Pig.

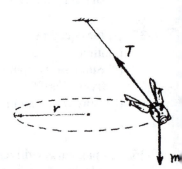

8-42. An airplane banks when going around a curve in a way similar to the banking of cars on circular tracks. The lifting force L on the plane plays the same role as the normal force N on the car.
(a) At what angle should a plane traveling at speed v make with the horizontal when the radius of the horizontal turn is r?
(b) What is the actual angle of banking if the speed of the airplane is 200.0 m/s and the radius is 9000.0 m?
(c) How will this angle differ for an airplane of twice the mass traveling at the same speed and making the same radial turn?

8-43. • At a popular carnival ride one sits in a chair that is swung in a
circular path by a cable as shown. The occupant, in effect, is the
bob of a conical pendulum. The mass of an occupant and the
chair is m, the length of the cable L, and the angle the cable
makes with the vertical is θ.

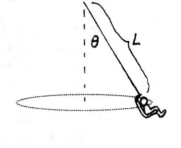

(a) What is the tension T in the cable?

(b) Calculate the tension if the mass of the occupant and chair is
185 kg, the length of the cable is 15.0 m and the angle the cable
makes with the vertical is 60°.

(c) Calculate the speed in the above situation.

8-44. • Whereas friction provides the centripetal force to hold a car on a
curved non-banked track, a properly-banked curve requires no
friction at all. Consider a horizontal circular track of radius r
banked at angle θ.

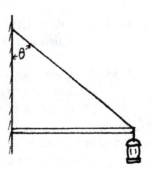

(a) Inspect the sketch and compare the vertical component of N
with the magnitude of mg. Why are they the same size?

(b) What role does the horizontal component of N play?

(c) Find the speed of a car of mass m that will enable it to remain
on the track without the benefit of friction.

(d) Calculate the speed of the car if the angle is 30°, radius is 125 m, and the mass of the
vehicle unknown.

8-45. • A car rounding a curve at speed v can make a turn on a banked turn without the aid of
friction. Consider a banked track of radius r. There are two principal forces acting on the
car; weight and the normal force.

(a) At what angle should the curve be banked so that cars can travel safely at speed v
without relying on friction?

(b) Calculate the banking angle for a curve of radius 180.0 m designed for a driving speed of
24.6 m/s.

8-46. • A lantern of weight W is suspended at the end of a
horizontal bar of weight w and length L that is supported by
a cable that makes an angle θ with the side of a vertical wall.
Assume the weight of the bar is at its center.

(a) Find the tension in the supporting cable.

(b) Calculate the tension in the cable for a bar of weight
28 N and length 1.5 m, plus a lantern of weight 85 N,
and the cable making a 37°-angle to the vertical.

8-47. • Consider a ball rolling at constant speed on the inner surface of a
cone. The surface of the cone makes an angle θ to the vertical as
shown. Further assume friction is negligible and the ball rolls in a
horizontal circle of radius r.

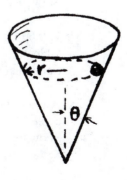

(a) Only two forces act on the ball. What are they?

(b) Is the normal force greater than, less than, or equal to the weight?

(c) Find the speed of the ball.

Show-That Problems

8-48. A child rides a flying horse at the outer edge of a merry-go-round. She is located 9.5 m from the central axis and is a bit frightened of the speed. So her parents place her on a horse nearer the center, 4.5 meters from the axis.
Show that the inner horse has a linear speed less than half the speed of the outer horse.

8-49. A 45-N force perpendicular to the 0.20-m-long handle of a wrench is successful in turning a stubborn bolt.
Show that the torque produced is 9.0 N·m, and that when the length of the handle is extended to 0.5 m with a length of pipe the torque is 2.5 times as much.

8-50. A boy weighs 388 N and sits 2.0 m from the fulcrum of a seesaw. The girl, on the other side of the fulcrum, sits 3.0 m from the fulcrum.
Show that the girl weighs 259 N.

8-51. A 388-N boy and a 259-N girl balance on a seesaw. The boy sits 2.0 m from the fulcrum of the seesaw and the girl sits on the other side, 3.0 m from the fulcrum. The girl is handed a bag of apples weighing 75 N.
Show that she will have to move 68 cm closer to the fulcrum for the seesaw to remain in balance.

8-52. A 3.0-kg toy is tied to a string. Stephanie holds one end of the string above her head and whirls the toy so that it makes a horizontal circle of radius 1.2 m at waist level. The toy is effectively the bob of a conical pendulum, moving at 3.4 m/s with a string tension of 41 N.
Show that the angle of the string with the vertical is about 45°.

8-53. Consider a car rounding a horizontal non-banked curve of radius 55 m.
Show that the coefficient of static friction between the road and a car's tires must be at least 0.60 if the car is to round the curve at a speed of 18 m/s.

8-54. Because of Earth's spin, your weight at the equator is slightly less.
Show that you would weigh 0.3% less at the equator than you would at the North Pole.

8-55. Gretchen moves at a speed of 3.2 m/s when sitting on the edge of a horizontal rotating platform of diameter 4.2 m. Her mass is 46 kg.
Show that the frictional force needed to keep her from slipping off is about 224 N.

8-56. Suppose that the speed doubles for Gretchen in the previous problem, and she is able to hold onto the platform.
Show that her angular momentum about the center of the platform will be 618 kg·m²/s.

8-57. A 0.60-kg puck revolves at 2.4 m/s at the end of a 0.90-m string on a frictionless air table.
Show that when the string is shortened to 0.60 m the speed of the puck will be 3.6 m/s.

8-58. An astronaut on a space walk swings a 1.2-kg bucket in a circular path at the end of a cable 3.6 m long. The astronaut then reels in the bucket to a 1.2-m distance.
Show that the speed of the bucket will be three times as great at the reduced distance.

8-59. An ice skater spins about a vertical axis with arms outstretched. She brings her arms in and reduces her rotational inertia to one-third her initial rotational inertia.
Show that she spins three times faster.

8-60. A skater rotates at 1.2 revolutions per second with arms at her side. When she raises her arms to a horizontal position, her speed decreases to 0.8 revolutions per second.
Show that her rotational inertia with outstretched arms has increased by 1.5 times.

9   *Gravity*

Just as beautiful music has a pattern that musicians of all nationalities comprehend, so a universal pattern underlies the cosmos. The stars, planets, and all other bodies in the universe dance in rhythm with this pattern—the universal law of gravitation. Isaac Newton discovered that every body in the universe is pulling on every other body with a force, $F = G\dfrac{m_1 m_2}{d^2}$. Here m_1 and m_2 are the masses of a pair of bodies and d is the distance between their centers. The force law as written applies to both particles and spherical bodies, as well as to non-spherical bodies sufficiently far apart.

    The problems in this chapter are applications of this central rule of nature—as expressed in the equation for gravitational force.

Sample Problem 1

A group of students repeat Philipp von Jolly's experiment (discussed on pages 163-164 of *Conceptual Physics, 10th Edition*). They record the following data: $m_1 = 5.00$ kg, $m_2 = 5{,}775$ kg, $F =$ the weight of 0.589 milligram (the *extra* mass needed to maintain balance after the large sphere is rolled under the smaller one), and $d = 0.569$ m, the distance between the centers of m_1 and m_2.

(a) With these data compute the universal gravitational constant, G.

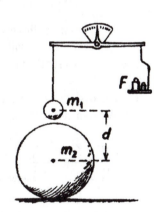

Focus: $G=?$

$$\text{From } F = G\frac{m_1 m_2}{d^2} \quad \Rightarrow G = \frac{F}{\frac{m_1 m_2}{d^2}} = \frac{Fd^2}{m_1 m_2}.$$

The difficulty in measuring G is that the gravitational force between ordinary bodies is so small! We can see just *how* small it is by recognizing that the pull between the 5.00 kg and the 5775 kg masses just over half a meter apart can be balanced out by Earth's pull on 0.589 milligrams (about the weight of the tiniest bit of aluminum foil you can tear from a roll of foil). Nevertheless, students in physics labs, using methods such as von Jolly's, can measure incredibly weak forces. The gravitational force between the spherical flask of mercury, m_1, and the big ball of lead, m_2, effectively equals the weight of 0.589 milligrams of material. Call the mass of this bit of matter m, and its weight mg.

$$\text{Then we get } G = \frac{Fd^2}{m_1 m_2} = \frac{mgd^2}{m_1 m_2} = \frac{\left(0.589\,\text{mg} \times \frac{1\,\text{g}}{1000\,\text{mg}} \times \frac{1\,\text{kg}}{1000\,\text{g}}\right)\left(9.8\,\tfrac{\text{m}}{\text{s}^2}\right)(0.569\,\text{m})^2}{(5.00\,\text{kg})(5775\,\text{kg})}$$

$$= 6.47 \times 10^{-11}\,\frac{\text{N} \cdot \text{m}^2}{\text{kg}^2}, \text{ about 3\% less than the value of } G \text{ accepted today.}$$

(b) If this experiment were performed on the Moon, which values would change and which wouldn't?

Answer: m_1, m_2, and d are independent of location so those values wouldn't change. Neither would the force between the two masses. But because g on the Moon is about six times smaller than it is on Earth, we'd have to add about six times as much extra mass onto the balance plate to get the same mg as occurs at Earth's surface.

Sample Problem 2

Sam, having read about a new "lose-weight-as-you-sleep" scheme, decides to move his bed upstairs, 3 m farther from the ground than he was before.

(a) How will his weight upstairs compare with his weight downstairs?

Focus: $\dfrac{W_{upstairs}}{W_{downstairs}} = ?$

Sam believes, correctly, that by moving his bed farther away from the center of Earth he will feel a smaller gravitational attraction to Earth—that he will "lose weight". The question is, how much weight will he lose?

His distance from Earth's center will increase from R_E (the radius of Earth) to $R_E + 3$ m. We can calculate his weight upstairs and downstairs using the law of universal gravitation:

$$\frac{W_{upstairs}}{W_{downstairs}} = \frac{F_{Earth\ and\ Sam\ upstairs}}{F_{Earth\ and\ Sam\ downstairs}} = \frac{G\dfrac{M_E m_s}{(R_E+3m)^2}}{G\dfrac{M_E m_s}{R_E^2}} = \frac{R_E^2}{(R_E+3m)^2} = \left(\frac{R_E}{R_E+3m}\right)^2$$

$$= \left(\frac{6,370,000\,m}{6,370,000\,m+3\,m}\right)^2 \approx \mathbf{0.999999999}$$

Sam's weight is reduced by 1 minus 0.999999, about one one-millionth. Of course he could achieve a similar weight loss by sweating the equivalent of about 2 drops of water (which he likely lost when he moved his bed upstairs!).

(b) Sam weighs 940 N downstairs. How much will he weigh upstairs?

$W_{upstairs} = 0.999999999\,W_{downstairs} = 0.999999999\,(940\,N)$ which basically $= \mathbf{940\,N}$.

No noticable change in weight.

Sample Problem 3

Your spaceship, mass m, is coasting at a point halfway between the Earth (mass m_E) and the Moon (mass m_M). The distance between the centers of Earth and the Moon is r_{EM}.

(a) Find the magnitude and direction of the net force on your ship.

> *Focus: $F_{net} = ?$*
>
> Your ship, with its rocket engine off, is subject to two forces, one pulling your ship toward the Moon (to the left in the sketch) and another pulling your ship toward Earth (to the right in the sketch). The distance between your ship and the center of either body is $r_{EM}/2$. The *net* force on your ship will be the vector sum of these two forces acting in opposite directions.
>
> If we call "toward Earth" the positive direction, then

$$F_{net} = F_{toward\ Earth} - F_{toward\ Moon} = G\frac{m_E m}{\left(\frac{r_{EM}}{2}\right)^2} - G\frac{m_M m}{\left(\frac{r_{EM}}{2}\right)^2} = G\frac{m}{\left(\frac{r_{EM}}{2}\right)^2}(m_E - m_M) = \frac{4Gm}{r_{EM}^2}(m_E - m_M).$$

Since you're the same distance from both Earth and the Moon, and Earth is the more massive of the two bodies, Earth's pull on your ship is greater than the Moon's. The net force points toward the Earth.

(b) Find the magnitude and direction of the acceleration of your 35,000-kg ship halfway between Earth and the Moon.

> *Focus: $a = ?$*

$$a = \frac{F_{net}}{m} = \frac{\left(\frac{4Gm}{r_{EM}^2}(m_E - m_M)\right)}{m} = \frac{4G}{r_{EM}^2}(m_E - m_M)$$

$$= \frac{4\left(6.67 \times 10^{-11} \text{N} \cdot \frac{\text{m}^2}{\text{kg}^2}\right)}{(3.84 \times 10^8 \text{m})^2} \cdot (5.98 \times 10^{24} \text{kg} - 7.36 \times 10^{22} \text{kg}) = 0.0107\,\tfrac{\text{m}}{\text{s}^2} = 1.07\,\tfrac{\text{cm}}{\text{s}^2}.$$

Your ship accelerates slowly toward Earth.

(c) If your ship were 10 times more massive, what would its acceleration be?

> *Answer:* $a = \dfrac{\text{ten times the force } F}{\text{ten times the mass } m} = a$.

A 10 times more massive ship means 10 times more gravitational force on it. So the acceleration of your ship would be the same (note that the mass of the ship cancels out in the solution for part b). The same cancellation of mass occurs for all freely falling bodies. In accord with Newton's second law, the ratio of force to mass for all bodies in free fall—including this coasting spaceship—is the same, whatever the location. That ratio is acceleration.

Gravity Problems

9-1. At Earth's surface, the gravitational field averages 9.80 N/kg. The field decreases with distance from Earth via the inverse-square law.

(a) At what distance from Earth's center is the gravitational field 0.098 N/kg? (Express your answer in Earth radii.)

(b) At what distance from Earth's *surface* is the gravitational field 0.098 N/kg?

(c) Show with units that $9.80\dfrac{\text{N}}{\text{kg}} = 9.80\dfrac{\text{m}}{\text{s}^2}$.

9-2. An object of mass m is brought to the Moon, where the acceleration due to gravity at its surface is only one-sixth that of Earth.

(a) What is the mass of the object on the Moon's surface?

(b) What is the weight of the object on the Moon's surface?

(c) If the object has a mass of 1.0 kg on Earth, what would be its mass and weight on the Moon?

9-3. Two asteroids with masses m and M are separated by a distance d and attract each other with a gravitational force F.

(a) By how much does the force change if both masses are doubled, while the distance between the asteroids remains unchanged?

(b) By how much does the force change if both masses double and the distance between them also doubles?

(c) By how much does the force change if only one mass doubles, and the distance between them doubles?

9-4. Suppose that the mass of a star increases by a factor of four while its diameter doubles.

(a) What is the change in the gravitational field at its surface?

(b) Another star of the same mass collapses to one-tenth its normal radius with no mass change. How is the gravitational field at its surface affected?

9-5. Venus has a mass 81.5% that of Earth, and its radius is 94.9% that of Earth.

(a) Calculate the value of g on the surface of Venus.

(b) Calculate the weight of a 1.0-kg stone on the surface of Venus.

9-6. The Hubble Space Telescope is 570 km above Earth's surface.

(a) What is Earth's force of gravity on an 85-kg astronaut on Earth's surface.

(b) Calculate Earth's force of gravity of the same astronaut doing repairs on the Hubble Telescope.

9-7. You (mass m_Y) see a charming someone (mass m_C) across the room a distance x away.

(a) Approximately what is the strength of your gravitational attraction to the charming person?

(b) What is the charming person's attraction to you?

(c) Calculate the strength of the attraction if you each have a mass of 75 kg and you stand 1.5 m apart.

9-8. A mass m is suspended from a vertical spring (recall Hooke's law: $F = kx$, where the stretch of a spring is proportional to the force on it). The spring stretches a distance x_E on Earth. The same mass on the surface of another planet stretches the spring by a distance x_P.
   (a) Find the value of g on the other planet.
   (b) Suppose that the planet were inflated in size but kept the same mass. At the new surface, would the spring stretch more, less, or the same amount as before? Defend your answer.

9-9. After you blast off in a rocket ship, the force of Earth gravity on you decreases. Earth's radius is R and its mass is M.
   (a) At what distance from Earth's center would you weigh one-quarter as much as you do at the Earth's surface? (Answer in Earth radii.)
   (b) At what distance from Earth's center, in Earth radii, would you weigh one-tenth your weight at Earth's surface?

9-10. The Earth (mass m_E) and the Moon (mass m_M) are separated by a distance R_{EM}. At some location between them their gravitational fields cancel.
   (a) At what location between the centers of Earth and the Moon would the net gravitational field due to the Earth and Moon be zero?
   (b) Using the actual masses of Earth and the Moon and the actual distance between their centers, find how far, in meters, this point is from the center of Earth and from the center of the Moon.

9-11. The International Space Station (ISS) orbits the Earth at a distance h above Earth's surface. The radius of the Earth is R_E.
   (a) Find the acceleration due to gravity at the location of the ISS.
   (b) In what direction would you have to propel a coin at the ISS so that it would fall straight down to Earth?

9-12. Pretend that the Sun (mass m_S and radius R_S) shrinks to a white dwarf with a radius equal to that of the Earth, R_E.
   (a) Find the acceleration due to gravity at the surface of the shrunken Sun.
   (b) If the Sun shrank still further, to half R_E, what would g be at its surface?

9-13. Neil Armstrong throws a ball straight up at an initial speed v from the surface of the Moon (mass m_M, radius R_M).
   (a) Write a formula for g at the Moon's surface.
   (b) What maximum height will the ball reach given this initial speed?
   (c) How much time after the ball leaves Neil's hand could he catch it again?

9-14. When you step on a weighing scale your downward weight W and the upward support force N (the normal force) in effect compress a spring that is calibrated to show your weight. What will your weight be in an elevator that
   (a) accelerates upward with acceleration a?
   (b) accelerates downward with acceleration a?
   (c) moves upward with a constant velocity v?
   (d) moves downward with a constant velocity v?

9-15. You, mass m, stand on a bathroom scale on the surface of a planet of radius R and find that your weight is W.

(a) Find the mass of the planet from this information.

(b) Calculate the planet's mass if your weight on the planet is 469 N, your mass is 67.0 kg, and the radius of the planet is 5.97×10^6 meters.

9-16. Planet X has a mass that is x times the mass of Earth and a radius y times the radius of Earth. Your weight on the surface of Earth is W_E.

(a) Write an expression for the ratio of your weight on Planet X compared with your weight on Earth.

(b) How much would your weight be on the surface of the same planet if its diameter shrunk to half its original size with no change in its mass?

9-17. On Earth's surface you have weight, W. Pretend that somehow the Earth shrank in size, keeping the same mass, with no effects due to Earth's rotation.

(a) With what force would gravity pull on you at the surface of a shrunken Earth if it shrinks to 80% of its present radius?

(b) With what force would gravity pull on you if Earth shrank to 10% its present radius?

(c) Instead of shrinking, pretend that you could burrow to Earth's center. What would your weight be there?

9-18. A blob of water on Earth is gravitationally attracted to the Sun, and also to the Moon. Data for a comparison can be found on the inside back cover of your textbook.

(a) Which pulls harder on the blob of water on Earth, the Sun, or the Moon?

(b) How many times larger is the strongest pull than the weakest pull?

(c) How do you reconcile your answers with the fact that the Moon is more responsible for ocean tides on Earth than the Sun is?

9-19. Consider two identical 1.0-kg blobs of water on opposite sides of Earth, one on the side facing the Moon and the other on the side farthest away from the Moon.

(a) Calculate the gravitational force of the Moon on the blob on the side of Earth closest to the Moon. (Information you'll need is on the inside back cover of your textbook. The Earth-Moon distance listed there is measured center-to-center.)

(b) Calculate the gravitational force of the Moon on the blob on the side of Earth farthest from the Moon.

(c) Calculate the percent difference between these two forces ($\frac{F_{\text{near blob}} - F_{\text{far blob}}}{F_{\text{near blob}}} \times 100\%$).

(d) How does this difference relate to the existence of tides?

9-20. The Sun has a mass 333,000 times that of Earth. While you stand on Earth the average distance between you and the center of the Sun is 23,500 times the distance between you and the Earth's center.

(a) Which pulls harder on you, the Sun or the Earth?

(b) Calculate the ratio of the two pulls on you.

9-21. Four identical asteroids of mass m occupy the corners of a large square in space. The sides of the square are L.

(a) What is the net gravitational force (magnitude and direction) on one of the asteroids, due to the other three? (Hint: These forces combine as vectors.)

(b) What is the net gravitational field at the center of the "square" due to the four asteroids?

Show-That Problems for Gravity

9-22. Equate your weight mg to Newton's equation for gravitational force, $G\dfrac{mM}{R^2}$, where M is the mass of Earth and R its radius.

Show that $g = \dfrac{GM}{R^2}$.

9-23. The symbol g can mean acceleration due to gravity or gravitational field strength. Show that its units can be m/s^2 or N/kg.

9-24. Aileen drops a dense mass m from a high tower and it undergoes an acceleration g. Show that if she dropped a dense mass $2m$ it would have the same acceleration.

9-25. Consider two identical twins, each of mass 40 kg located 1 meter apart. Show that the gravitational attraction between them is the same as the weight of a 10-microgram speck of dust. (1 microgram $= 10^{-6}$ grams.)

9-26. Suppose that the mass of a star somehow doubles without any change in its radius. Show that the acceleration due to gravity at its surface also doubles.

9-27. Suppose that the diameter of a star doubles, without a change in mass. Show that the acceleration due to gravity at its surface reduces to 1/4 its previous value.

9-28. In a high-flying plane you are farther from Earth's center than at the surface. Show that at a 10-km altitude gravity's pull on you is $\sim 0.3\%$ less than it is on the ground.

9-29. The International Space Station orbits about 360 km above the Earth's surface. Show that the acceleration due to Earth's gravity at this distance is 90% of the value at Earth's surface.

9-30. Jack climbs a beanstalk so high that his weight at the top is half that at Earth's surface. Show that Jack is 9010 km from Earth's center.

9-31. The Moon is being pulled by both the nearby Earth and the faraway Sun. Show that the Sun pulls on the Moon with more than twice as much force as the Earth does.

9-32. Ocean tides are caused more by the Moon than the Sun. This might imply that the closeness of the Moon results in a greater pull on our oceans. But such is not the case. Show that gravitation between Earth and the Sun is 177 times stronger than gravitation between Earth and the Moon.

9-33. A 1.0-kg melon weighs 9.8 N at Earth's surface. If the melon is located two Earth radii above the Earth, its weight would be less.
Show that the gravitational force on the melon would be 1.1 N at this distance.

9-34. Consider a mountain of mass 6.0×10^{11} kg, 1.0 km away from you.
Show that the mountain exerts more gravitational force on you than the Moon.
(Moon's mass is 7.4×10^{22} kg, located 3.8×10^5 km away).

9-35. A newborn baby of mass 3.5 kg is held in the arms of her mother who has a mass 75 kg. Assume the distance between their centers of gravity is 0.30 m.
Show that the force of gravity between mother and child is more than eight times as strong as the gravitational force exerted by Mars on the child.
(The mass of Mars is 6.4×10^{23} kg and its closest distance to Earth is 8.0×10^{10} m.)

The central idea in solving projectile motion problems is recognizing that the horizontal and vertical components of the motion are independent. In the absence of air resistance the only force acting on a projectile is gravitational, so the only acceleration of the projectile is downward. The horizontal motion proceeds as though the vertical motion wasn't there.

Some of the projectile problems in this section have a bit of trig, and some employ momentum and/or energy concepts. Equating gravitational force to centripetal force is often a good route to solving problems involving satellites in circular orbit. Some problems involve the conservation of angular momentum and some rely on Kepler's laws of planetary motion. It is fitting that this chapter ends our treatment of mechanics.

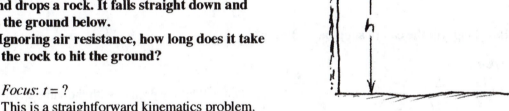

Sample Problem 1
Zephram leans over the edge of a cliff of height
***h* and drops a rock. It falls straight down and**
hits the ground below.
(a) Ignoring air resistance, how long does it take
    the rock to hit the ground?

Focus: t = ?
This is a straightforward kinematics problem.

From $d = v_0 t + \frac{1}{2} a t^2$ with $v_0 = 0, d = h$ and $a = g, \ d = \frac{1}{2} a t^2 \ \Rightarrow t = \sqrt{\dfrac{2h}{g}}.$

(b) Suppose that the rock is instead thrown horizontally from the cliff at a speed v_x. How
    long is the rock in the air before it hits the ground?

Focus: t = ?
Since the horizontal motion of the rock doesn't affect its vertical motion, it will still hit

the ground in a time $t = \sqrt{\dfrac{2h}{g}}.$

(c) How far from the base of the cliff will the rock land?

Focus: x = ?

From $x = v_x t = v_x \sqrt{\dfrac{2h}{g}}.$ The fact that the rock is falling doesn't affect its horizontal velocity.

(d) Calculate how far from the base of the cliff the rock lands when it is thrown horizontally
    at 12 m/s. The cliff is 33 m high.

Solution: $x = v_x \sqrt{\dfrac{2h}{g}} = 12\tfrac{m}{s} \sqrt{\dfrac{2(33 \text{ m})}{\left(9.8\tfrac{m}{s^2}\right)}} = \textbf{31 m.}$

Sample Problem 2

A soup can of mass m is projected at a speed v from the ground at an angle θ with the ground.

(a) What are the horizontal and vertical components of the can's velocity?

From the diagram, $\cos\theta = \dfrac{v_x}{v} \Rightarrow v_x = v\cos\theta$.

Similarly, $\sin\theta = \dfrac{v_y}{v} \Rightarrow v_y = v\sin\theta$.

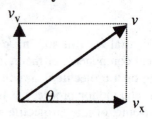

(b) How high above the ground will the soup can rise?

Focus: h = ?

Since we're not given the time, we'll apply the time-independent kinematics equation $v_f^2 - v_0^2 = 2ad$ to the vertical direction. At maximum height h, $v_f = 0$, so the equation reduces to $-v_0^2 = 2ah$. Initial speed v_0 in the vertical direction is $v\sin\theta$ and acceleration a is $-g$. So the equation becomes $-(v\sin\theta)^2 = 2(-g)h$

$$\Rightarrow h = \frac{(v\sin\theta)^2}{2g}.$$

(c) How long will the can of soup be in the air?

Focus: t = ?

From $a = \dfrac{\Delta v}{\Delta t} = \dfrac{v_f - v_0}{t} \Rightarrow t = \dfrac{v_f - v_0}{a}$. The can will hit the ground with the same speed v at which it was launched. When hitting, the y-component of its velocity will be $-v\sin\theta$.

$$\text{So } t = \frac{v_f - v_0}{a} = \frac{-v\sin\theta - v\sin\theta}{-g} = \frac{2v\sin\theta}{g}.$$

Another approach is to find how long the soup can takes to reach its maximum height, and then double that time to find the time for the whole trip.

From $v_f = v_0 + at \Rightarrow 0 = v\sin\theta + (-g)t \Rightarrow t = \dfrac{v\sin\theta}{g}$. This is the time for the can to reach

maximum height. The total time in the air is twice this $\Rightarrow t = \dfrac{2v\sin\theta}{g}$.

(d) How far away from the launch point will the can hit the ground?

Focus: x=?

$x = v_x t = (v\cos\theta)t$. The vertical motion of the can doesn't affect its horizontal motion. All we need here is the time in the air, which we found in part (c). Then

$$x = v_x t = (v\cos\theta)\frac{2v\sin\theta}{g} = \frac{2v^2\sin\theta\cos\theta}{g}.$$

(e) For a given initial velocity v, what throwing angle will give the maximum range for the soup can?

Focus: x_{max}=?

For a given v and value of g the maximum horizontal range will depend only upon the initial angle θ. If we make a table we can see that the maximum range occurs when the launch angle is 45°.

θ	$\sin\theta$	$\cos\theta$	$\sin\theta \times \cos\theta$
15°	0.259	0.966	0.250
30°	0.500	0.866	0.433
40°	0.643	0.766	0.492
45°	0.707	0.707	0.500
50°	0.766	0.643	0.492
60°	0.866	0.500	0.433
75°	0.966	0.259	0.250

(f) Calculate the maximum height, time in the air, and horizontal range for a 0.4-kg soup can thrown from the ground at 21 m/s at an angle of 35° above the horizontal.

Solution:

$$h = \frac{(v\sin\theta)^2}{2g} = \frac{\left(21\frac{m}{s}\sin35°\right)^2}{2\left(9.8\frac{m}{s^2}\right)} = 7.4 \text{ m.}$$

$$t = \frac{2v\sin\theta}{g} = \frac{2\left(21\frac{m}{s}\right)\sin35°}{9.8\frac{m}{s^2}} = 2.5 \text{ s.}$$

$$x = \frac{2v^2\sin\theta\cos\theta}{g} = \frac{2\left(21\frac{m}{s}\right)^2\sin35°\cos35°}{9.8\frac{m}{s^2}} = 42 \text{ m.}$$

Sample Problem 3

Consider a moving tennis ball that, at the highest point in its path just barely clears the net, a distance y above the ground. To remain within the tennis court it can't be moving too fast.

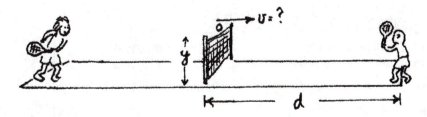

(a) Ignoring air drag, what is the maximum speed the ball can have as it clears the net to strike within the court's border a distance d from the net?

Focus: $v = ?$

The physics concept here involves projectile motion in the absence of air drag, where horizontal and vertical components of velocity are independent. We're asked for horizontal speed, so we write,

$$v_x = \frac{d}{t}$$

where d is horizontal distance traveled in time t. We're given d but we're not given the time. So this isn't a simple one-step problem. Importantly, *the equation directs us to consider time*, which we might not have considered otherwise!

Now some important physics: The time t it takes for the ball to hit the court will be the same as if we had just dropped it from the top of the net from rest, a vertical distance y. We say from rest because the initial vertical component of velocity is zero—at the highest point in its path the ball moves horizontally over the net.

From $y = \frac{1}{2}gt^2 \Rightarrow t^2 = \frac{2y}{g} \Rightarrow t = \sqrt{\frac{2y}{g}}$.

So $v = \frac{d}{t} = \frac{d}{\sqrt{\dfrac{2y}{g}}}$.

(b) Calculate the maximum speed a horizontally-moving tennis ball can have as it clears the net of height 1.00 m and still strike within the court's border, 12.0 m distant from the net?

Solution: $v = \dfrac{d}{\sqrt{\dfrac{2y}{g}}} = \dfrac{12.0 \text{ m}}{\sqrt{\dfrac{2(1.00 \text{ m})}{9.80 \frac{\text{m}}{\text{s}^2}}}} = 26.6$ **m / s.**

Sample Problem 4

A watermelon rolls off the horizontal edge of a building at speed v.
(a) What is its speed after time t as it sails through the air? Ignore air resistance.

Focus: $v = ?$

At any given time there are two components of the watermelon's velocity—a vertical component v_y, and a horizontal component v_x. The speed at any moment is the magnitude of the vector sum of these two components: $speed = \sqrt{v_x^2 + v_y^2}$. The equation tells us to find expressions for v_x and v_y.

The horizontal component $v_x = v$, since no horizontal forces act on the watermelon to produce horizontal accelerations. The vertical component at any time t is $v_y = gt$.

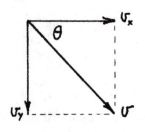

So $speed = \sqrt{v_x^2 + v_y^2} = \sqrt{v^2 + (gt)^2}$.

(b) What angle does it make with the horizontal at time t?

Focus; $\theta = ?$

Look at the diagram. v_x, v_y, and v form a right triangle, with v_y as the side opposite θ, v_x as the side adjacent to θ, with v as the hypotenuse. We can see that

$$\tan\theta = \frac{v_y}{v_x} \Rightarrow \theta = \tan^{-1}\left(\frac{v_y}{v_x}\right) = \tan^{-1}\left(\frac{gt}{v}\right).$$

(c) Calculate the velocity of the watermelon 1.0 second after rolling off the edge at speed 9.8 m/s.

Focus: $v = ?$

The question is asking us to find the speed **and** direction of the watermelon 1.0 second after rolling off the edge of the building.

$$speed = \sqrt{v^2 + (gt)^2} = \sqrt{\left(9.8\tfrac{m}{s}\right)^2 + \left(9.8\tfrac{m}{s^2}(1.0s)\right)^2} = 14\tfrac{m}{s};$$

$$\theta = \tan^{-1}\left(\frac{v_y}{v_x}\right) = \tan^{-1}\left(\frac{9.8\tfrac{m}{s^2}(1.0s)}{9.8\tfrac{m}{s}}\right) = 45°.$$

So the velocity of the watermelon at $t = 1.0$ s is 14 m/s at an angle of 45° below the horizontal. Could this have been solved without computations? Yes, for we see that the vertical component of velocity in 1 s equals the horizontal component. So that's 45° and $\sqrt{2}$ the original speed!

Sample Problem 5
Ramon fires a projectile at an angle θ above the horizontal ground at velocity v; the projectile strikes a balloon when it reaches the top of its trajectory.
(a) Ignoring air drag, what is the speed with which it hits the balloon?

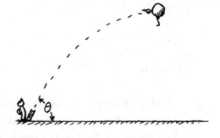

Focus: $v = ?$

At its maximum height the vertical component of the projectile's velocity is zero, so the magnitude of v will be the magnitude of the x-component of v_0: $v = v_{0x} = v_0\cos\theta$.

(b) Calculate the speed of the projectile when it hits the balloon. Assume its initial speed was 100.0 m/s and the projection angle 60°.

$$v = v_0\cos\theta = \left(100.0\tfrac{m}{s}\right)\cos60° = 50.0\tfrac{m}{s}.$$

(c) How high is the balloon?

Focus: $h = ?$

Since we're not given time, we consider the time-independent kinematics equation $v_f^2 - v_0^2 = 2ad$ applied to vertical motion, where a is $-g$ and d is h.

From $v_f^2 - v_0^2 = 2(-g)h$, $\quad h = \dfrac{-v_{0y}^2}{2(-g)} = \dfrac{(v_0\sin\theta)^2}{2g} = \dfrac{\left(100.0\tfrac{m}{s}\sin60°\right)^2}{2\left(9.80\tfrac{m}{s^2}\right)} = 383$ m. .

(d) The Sun is directly overhead and casts a shadow of the projectile on the ground as it sails through the air. How fast does the shadow move across the ground?

Answer: The shadow across the ground has the same speed as the horizontal component of the projectile's velocity. So $v = v_x = v_0 \cos\theta = \left(100.0 \tfrac{m}{s}\right)\cos 60° = \mathbf{50.0 \tfrac{m}{s}}$.

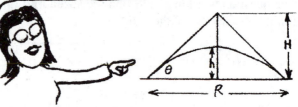

Interesting tidbit: The maximum height of a projectile following a parabolic path is half the height of an isosceles triangle whose base is equal to the range, with the angle of the two sides equal to the projectile's launch angle.

Sample Problem 6

A satellite of mass m circles the Earth at a constant altitude h above the Earth's surface.
(a) What is the speed of the satellite?

Focus: $v = ?$
Since the satellite travels in a circular orbit, the gravitational force must provide a centripetal force mv^2/r. Math-wise

$$G\frac{M_E m_{sat}}{\left(d_{Earth\,to\,sat}\right)^2} = \frac{m_{sat}v^2}{r_{orbit}}.$$

Three things to notice before we go any further:
- The term $d_{Earth\,to\,sat}$ is the distance from the *center* of the Earth to the satellite, equal to $R_E\,+$
- The radius of the orbit (r in the mv^2/r term) is also measured from the center of the Earth, and is exactly the same as $d_{Earth\,to\,sat}$.
- We don't need to know the mass of the satellite (because it cancels out of the equation).

From $G\dfrac{M_E m_{sat}}{\left(d_{Earth\,to\,sat}\right)^2} = \dfrac{m_{sat}v^2}{r_{orbit}}$ $\Rightarrow$ $G\dfrac{M_E}{\left(R_E+h\right)^2} = \dfrac{v^2}{\left(R_E+h\right)}$ $\Rightarrow v = \sqrt{\dfrac{GM_E}{\left(R_E+h\right)}}$.

(b) The International Space Station orbits Earth in a nearly circular orbit, averaging a height of 370 km above Earth's surface. What is the speed of the ISS as it circles Earth?

From above, $v = \sqrt{\dfrac{GM_E}{\left(R_E+h\right)}}$.

Two things to notice before we go any further:
- G has units of $N \cdot m^2/kg^2$, so all of the distances we use should be expressed in meters before we can use the equation.
- The mass and radius of Earth are listed in the inside back cover of your textbook!

First let's convert the units:

height $h = 370$ km $= 3.7 \times 10^5$ m.

$R_E + h = 6.37 \times 10^6$ m $+ 0.37 \times 10^6$ m $= 6.74 \times 10^6$ m .

Then $v = \sqrt{\dfrac{6.67 \times 10^{-11} \frac{N \cdot m^2}{kg^2}\left(5.98 \times 10^{24} kg\right)}{6.74 \times 10^6 m}} = \mathbf{7690\,\frac{m}{s}}$ (which is a brisk $27{,}700\,\frac{km}{h}$!).

The space station moves about 25 to 30 times faster than a normal commercial jet. This high speed ensures that it falls around and around Earth, and not into its surface!

(c) The Moon orbits Earth once each 27.3 days. Use this information to calculate the time that the International Space Station takes to complete an orbit.

T=? We can take two approaches:

Approach 1: From $v = \dfrac{d}{t} \Rightarrow t = \dfrac{2\pi r}{v} = \dfrac{2\pi(6.74 \times 10^6 m)}{7690\,\frac{m}{s}} = 5504 s \times \dfrac{1 h}{3600 s} = \mathbf{1.53\ h}$.

Approach 2: Both the space station and the Moon orbit Earth. From Kepler's 3rd Law, the (period)^2 of a satellite's orbit is proportional to its (orbital radius)^3.

$T_{orbit}^2 \sim r_{orbit}^3$ or $\left(\dfrac{T_{ISS}}{T_{Moon}}\right)^2 = \left(\dfrac{r_{ISS\,orbit}}{r_{Moon\,orbit}}\right)^3 \Rightarrow T_{ISS} = T_{Moon}\sqrt{\left(\dfrac{r_{ISS}}{r_{Moon}}\right)^3} = 27.3\,\text{days}\sqrt{\left(\dfrac{6.74 \times 10^6 m}{3.84 \times 10^8 m}\right)^3}$

$= 0.0635\,\text{days} \times \dfrac{24 h}{1 day} = \mathbf{1.52\ h}$, about the same answer obtained by the other approach.

Projectile and Satellite Motion Problems

10-1. Three balls are tossed off the roof of a building. Ball A is thrown straight downward with speed v. Ball B is thrown upward with speed v. Ball C is simply dropped from rest. Neglect air resistance.
(a) Which ball hits the ground first? Last?
(b) Which ball hits the ground with the greatest speed? The least speed?
(c) Which ball spends the most time in the air? The least time?

10-2. Three ballplayers, Abe, Bob, and Cal, throw balls straight up. Abe throws at speed v, Bob throws at speed $1.5v$, and Cal throws at speed $2v$. The balls each land at the same height from which they were thrown. Neglect air resistance.
(a) Which player's ball is in the air longest? Shortest?
(b) Which player's ball goes highest? Lowest?
(c) Which player's ball has the greatest acceleration in flight? The least acceleration?

10-3. Three projectiles are launched with the same initial velocity v. Projectile A is launched at angle of 30° from the horizontal, Projectile B at 45°, and Projectile C at 60°. Each projectile lands at the same elevation from which it was launched. Neglect air resistance.
(a) Which projectile has the greatest horizontal velocity? The least?
(b) Which projectile has the greatest initial vertical velocity? The least?
(c) Which projectile attains the greatest height? The least?
(d) Which projectile has the greatest range? The least?
(e) Which projectile has the greatest time in the air? The least?

10-4. A lemming waddles horizontally off a small cliff of height h with a speed v.
   (a) How far from the base of the cliff will the lemming land?
   (b) What will be the lemming's speed upon landing?
   (c) Calculate answers to parts (a) and (b) if the height of the cliff is 4.9 m, initial speed of the lemming is 0.50 m/s, and its mass is 0.4 kg.

10-5. Firemen shoot a stream of water at a burning building. Water leaves the high-pressure hose at speed v at an angle θ to the horizontal.
   (a) What is the speed of the water at the highest point in its trajectory? Ignore air resistance.
   (b) What is the acceleration of the water at its highest point? Again, ignore air resistance.

10-6. That strange classmate of yours slides your physics textbook of mass m off the lab table with a speed v. It meets the floor in time t.
   (a) How high above the floor is the tabletop?
   (b) How far horizontally from the edge of the table does the book land?
   (c) Calculate the answers to parts (a) and (b) if the book leaves the table at 1.2 m/s, the time in the air is 0.43 s, and you don't know the mass of the book.

10-7. Chuck gives a horizontal speed v to a ball that then rolls off a lab bench y meters high.
   (a) How long a time will it take the ball to reach the floor?
   (b) How far from a point on the floor directly below the edge of the bench will the ball land?
   (c) Calculate how long and how far for $v = 1.5$ m/s and a bench height of 1.2 m.

10-8. Chuck's ball rolls horizontally at speed v off the edge of a platform and lands a horizontal distance x beyond the edge of the platform.
   (a) How long is the ball in the air?
   (b) What is the height of a platform?
   (c) Calculate the height of the platform if the ball rolls off the edge at 15 m/s and lands 18 m away.

10-9. Manuel throws a ball horizontally with speed v from the top of a building of height h.
   (a) What horizontal distance from the bottom of the building does the ball strike the ground?
   (b) Calculate the horizontal distance from the bottom of the building that the ball will land for a horizontal speed 15 m/s and a building 32 m tall.
   (c) If the above situation were to take place on the surface of the Moon where the acceleration due to gravity is only $1/6\ g$, calculate the horizontal distance from the base of the lunar building that the ball lands.

10-10. Caillen stands atop a cliff and tosses a stone horizontally into the ocean below. The stone has a mass m, and hits the water in time t after leaving her hand.
   (a) How high is the cliff above the water?
   (b) Calculate the height of the cliff if the stone's mass is 0.80 kg and it hits the water 2.4 seconds after leaving Caillen's hand.

10-11. A bullet fired horizontally from the top of a tall building of height h has a muzzle
velocity v. At the same time a similar bullet is simply dropped from the same height and
takes time t to reach the ground below. Neglect air resistance.
(a) What is the horizontal range of the fired bullet?
(b) Calculate the range if the bullet's muzzle velocity is 310 m/s, the height of the
building is 19.6 m, and the time for the other bullet to drop is 2.0 s.

10-12. An emergency package of mass m is dropped from a horizontally-moving airplane
traveling at constant speed v when its altitude is h.
(a) How long does it take for the package to reach the ground, assuming no air drag?
(b) How far forward from its dropping point does the package land?
(c) At the moment the package hits the ground, what is its distance from the airplane?
(d) Calculate answers to these questions for the case of the airplane flying at 140 m/s
when the altitude is 0.50 km and the mass of the dropped package is 22 kg.

10-13. • Mala, practicing with her bow and arrow, fires one that strikes horizontally into a
third-floor barn window shutter. The arrow travels a horizontal distance x and a vertical
distance y in time t. (Note again: The arrow strikes horizontally into the window shutter.
What does this tell you about the vertical component of the arrow's velocity when it
strikes?)
(a) What was the speed of Mala's arrow when it hit the shutter?
(b) Calculate the speed of Mala's arrow when it left her bow if the horizontal distance
traveled is 32 m, the vertical distance is 9.5 m, and time of flight is 1.39 s.
(c) Could you have solved this problem if the time of flight was not given? Explain.

10-14. A pellet is fired from a rifle at a speed v at an angle θ above the horizontal.
(a) What maximum height does the pellet reach?
(b) How long does the pellet take to reach its maximum height?
(c) What is the horizontal range of the pellet?
(d) Calculate answers to the above questions for an initial speed of 150 m/s and an angle
of 37° above the horizontal.

10-15. Fred throws a ball at an angle θ above the horizontal with speed v. Neglect air drag.
(a) What is the ball's speed when it reaches its highest point?
(b) What is the ball's speed 1.0 s after reaching its highest point, assuming it is still in
the air?
(c) Calculate answers to parts (a) and (b) if the initial speed is 30 m/s and the angle
is 60° above the horizontal.

10-16. On a baseball field, the distance between the pitcher's mound and home plate is x.
Pedro pitches a fastball that moves horizontally as it leaves his hand and takes 0.45 s
to reach the home plate. Ignore air drag.
(a) What is the initial speed of the pitch?
(b) Calculate the speed for a distance of 18.4 m between the pitcher's mound and home
plate.
(c) Calculate the vertical drop of this ball on its way from the pitcher's mound to home
plate.
(d) What is the vertical drop of the ball when halfway to the plate?

10-17. In a new twist on birthday party games, a piñata hangs from the branch of a very tall tree. Marcus aims and fires a small coconut from the ground at an angle θ above the horizontal. He fires it at velocity v_0. The coconut strikes the piñata just as it reaches the top of its trajectory.

(a) Ignoring air drag, what is the speed with which the coconut hits the piñata?

(b) Calculate the speed of the coconut when it hits the piñata. Assume an initial speed of 18.0 m/s and a projection angle of 55°.

(c) How high is the piñata?

(d) How long will it take the candies that fall from the piñata to hit the ground?

10-18. Stranded on a tropical island, Roberta loads her slingshot and fires a compact seashell at a coconut. The seashell leaves the slingshot at ground level at a speed v, making an angle θ with the horizontal. The coconut is hit at the top of the shell's trajectory.

(a) Ignoring air drag, what is the speed with which the seashell hits the coconut?

(b) Calculate the speed of the seashell when it hits the coconut. Assume an initial speed of 24.0 m/s and a projection angle of 53°.

(c) How high is the coconut?

10-19. Stan the stuntman drops from a helicopter that is flying at a horizontal velocity v.

(a) What is Stan's initial velocity when he drops from the helicopter?

(b) What is Stan's velocity t seconds later?

(c) How long will it take Stan to hit the safety net a vertical distance y below?

(d) Find Stan's velocity when he hits the net and his time in the air if the helicopter's velocity is 12 m/s and it is 26 m above the safety net when Stan begins his drop.

10-20. • Stan the stuntman again drops from a helicopter that this time is flying at velocity v slightly downward at an angle θ with the horizontal.

(a) What is Stan's initial velocity when he drops from the helicopter?

(b) What is Stan's velocity t seconds later?

(c) How long will it take Stan to hit the safety net a vertical distance y below?

(d) Find Stan's velocity when he lands on the net and his time in the air if the helicopter's velocity is 12 m/s at an angle of 15° below the horizontal, and is 26 m above the safety net when Stan begins his drop.

10-21. Travis throws a football with an initial speed v_0 at an initial angle θ to the horizontal. Assume air drag is negligible.

(a) Find the time t at which the ball reaches its maximum height.

(b) Calculate the time for the ball to reach its maximum height if the initial speed is 14 m/s at an angle of 37°.

(c) Calculate the speed of the ball at its maximum height.

10-22. • Stephanie is lying on the ground outside, playing with a high-pressure water hose, squirting water in all directions. She finds when she's done playing that the ground is wet up to a distance d away from where she was standing with the hose.

(a) How fast was the water coming out of the hose?

(b) Calculate the speed of the water if the radius of the wet circle is 45 m.

10-23. • A projectile of mass m is fired at an angle θ above the horizontal with a speed v.
    (a) Find the horizontal and vertical components of the projectile's initial momentum.
    (b) What will be the horizontal and vertical components of the projectile's momentum when it reaches its highest point, assuming no air drag?
    (c) The projectile explodes into two equally-massive fragments at the top of its trajectory. One fragment, the first, falls vertically to the ground below. What is the speed of the second fragment immediately after the explosion?
    (d) How much farther away from the launch point will the second fragment land compared with where the projectile would have landed if it didn't explode?

10-24. The speed of a satellite in circular orbit about Earth depends on its distance from Earth's center. The force of gravity provides the centripetal force necessary for circular orbit.
    (a) Equate centripetal force to gravitational force and solve for the speed of a satellite circling at a distance r from Earth's center.
    (b) Show that this speed in close orbit, a distance just a bit greater than Earth's radius, is about 8 km/s. (Earth's mass and radius are on the inside back cover of your textbook.)
    (c) Show that the time it takes for the satellite to complete one orbit at this distance is a little less than 90 minutes.

10-25. The speed of a spaceship in circular orbit about the Moon depends on its distance above Moon's surface. (Let Moon's radius be R and its mass M.)
    (a) What is the speed for a satellite in a circular orbit at altitude h above the Moon's surface (that's a total radial distance $R + h$)?
    (b) How much time does the satellite take to complete a lunar orbit at this altitude?
    (c) Calculate the speed and period of the spaceship if its distance above Moon's surface is 25.0 km.
    (d) Why is there no statement about neglecting air drag with this problem?

10-26. Robert is calculating the orbit for a weather satellite so that it will circle the Earth with an orbital period T.
    (a) Find the radius of this orbit in terms of T.
    (b) How high above Earth's surface will the satellite orbit?
    (c) How fast will the satellite move in its orbit?
    (d) Calculate the altitude and speed of a weather satellite that circles the Earth every 150 minutes.

10-27. A rocket is fired from Earth's surface to a height equal to Earth's radius.
    (a) How does the force of gravity on the rocket at this height compare with the force of gravity on the rocket back at Earth's surface?
    (b) What speed is needed to put the rocket in a circular orbit at this height?

10-28. Satellites that transmit TV signals are in geosynchronous orbit (above the Earth's equator with an orbital period of 24 hours).
    (a) How high above the equator are these satellites?
    (b) Why is it especially nice to have a satellite with a period of 24 hours? (Think about how satellite dishes are oriented when installing satellite TV.)

10-29. A satellite at distance r from the center of the Earth completes one circular orbit in time T. Newton's law of gravitation gives the pull of Earth's gravity on the satellite. This same pull is given by the equation for centripetal force.

(a) Equate gravitational force to centripetal force and solve for the mass M of the Earth in terms of r, T, and the gravitational constant G.

(b) Show that Kepler's Third Law of Planetary Motion follows from this relationship among M, r, T, and G.

10-30. The radius of Earth's orbit around the Sun is 1.5×10^{11}m. The period of Earth's orbit around the Sun is 365.25 days.

(a) Use this information and Kepler's Third Law to compute the mass of the Sun.

(b) Why does this information and Kepler's Third Law not allow you to calculate the mass of Earth?

(c) Why can the masses of planetary bodies with satellites be easily found, but not the masses of bodies without satellites?

10-31. Io, a satellite of Jupiter, has a mean distance from Jupiter of 4.22×10^8 m and a period of 1.77 Earth days.

(a) Use this information and Kepler's Third Law to compute the mass of Jupiter.

(b) Why cannot this information and Kepler's Third Law allow you to calculate the mass of Io?

10-32. You are exploring an as-yet uncharted solar system and you discover a small planet orbiting its parent star in a circular orbit of radius r. The planet has an orbital period T.
(a) What is the mass of this star?
(b) The mass of the Moon was unknown before 1959. Why?

10-33. An astronomical unit (AU) is defined as the average distance between the centers of the Earth and the Sun. Jupiter is on average 5.19 AU from the Sun.
(a) What is Jupiter's orbital period around the Sun?
(b) How would Jupiter's orbital period be affected if Jupiter were closer to the Sun?

Show-That Problems

10-34. The speed of a projectile anywhere along its path is given by $V = \sqrt{v_x^2 + v_y^2}$.

Show that the speed of a projectile at the peak of its trajectory is $V\cos\theta$, where θ is the angle V makes with the horizontal.

10-35. A waterfall flowing over the edge of a 28-m high cliff lands 5.0 meters from the base of the cliff.

Show that the speed of the water at the top of the cliff is 2.1 m/s.

10-36. Suppose that a flight attendant on a commercial jet moving 840 km/h pours a cup of coffee from a pot held 28 cm above the coffee cup.

Show that the coffee travels a horizontal distance of 56 meters relative to the Earth between the time it leaves the pot and the time it lands in the cup.

10-37. Tenny's compound archery bow can fire an arrow at 55 m/s (or more). Assume the arrow is shot horizontally at 55 m/s, pointed directly at the bull's-eye of a target 22.6 m away.

Show that the arrow hits the target 83 cm below the bull's-eye.

10-38. Suppose Tenny fires the same arrow as in the preceding problem and aims it 2.1° above the horizontal.

Show that she will hit the bull's eye.

10-39. • Diane can throw something up at a particular angle from the ground so that its maximum height and its horizontal distance traveled will be the same.

Show that this angle is 76°.

10-40. On February 6, 1971, Apollo 14 astronaut Alan Shepard hit a golf ball on the Moon with an improvised golf club he is rumored to have smuggled aboard the ship. The acceleration due to gravity on the surface of the Moon is 1.62 m/s².

Show that the ball would land about 480 meters away if he gave it an initial velocity of 32 m/s at an angle of 25° with the horizontal ground.

10-41. Equate the force of gravity to centripetal force where M is the mass of Earth and G the gravitational constant.

Show that the speed of a satellite in circular Earth orbit of radius r is $\sqrt{\dfrac{GM}{r}}$.

10-42. The distance covered in one complete circular orbit is simply the circumference of a circle.

Show that the speed of a satellite in circular orbit is $\dfrac{2\pi r}{T}$ where r is the radius of the satellite's orbit and T its orbital period.

10-43. Combine the results of the previous two problems (10-41 and 10-42).

Show that the period $T = 2\pi \sqrt{\dfrac{r^3}{GM}}$ for a satellite in circular orbit about Earth.

10-44. Titan is a moon of Saturn that orbits Saturn at an average distance of 1.22 million km with an orbital period of 15.9 days.

Show that the mass of Saturn is 5.69×10^{26} kg.

10-45. Mars has two moons, Phobos and Deimos. The radius of Mars is 3,397 km. Phobos orbits Mars at a distance of 5,980 km above Mars' surface with a period of 0.319 days. Deimos orbits Mars with a period of 1.262 days.

Show that Deimos is approximately 20,100 km above the surface of Mars.

10-46. Consider the data in the previous question.

Show that Phobos moves 1.58 times faster than Deimos if their orbits are approximated as circles.

10-47. Continue with the data in problem 10-45.

Show that the mass of Mars is 6.4×10^{23} kg.

10-48. A particular satellite circles Earth at a distance of 775 km above the surface.

Show that the orbital speed of the satellite is 7,470 m/s.

10-49. Another satellite orbits Earth at 4,990 m/s in a circular orbit.

Show that its distance from Earth's center is 16,000 km.

10-50. Make a simple force-vector diagram of a satellite in a circular orbit.

Show that Earth's gravitational force does no work on the satellite.

Although scaling is the highlight of this chapter, the concept of density is introduced here and is then covered in greater depth in Chapter 13.

Sample Problem 1
A movie depicts two pirates carrying a chest full of pure gold, which has a density of 19,300 kg/m³.

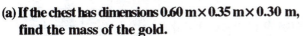

(a) If the chest has dimensions 0.60 m × 0.35 m × 0.30 m, find the mass of the gold.

Focus: $m = ?$ From the dimensions of the chest we can get its volume. From the density we can figure out the mass of that volume of gold.

From $\rho = \dfrac{m}{V} \;\Rightarrow\; m = \rho V = \rho(l \times w \times h) = 19{,}300\,\tfrac{\text{kg}}{\text{m}^3}(0.60\,\text{m} \times 0.35\,\text{m} \times 0.30\,\text{m}) = \mathbf{1{,}200\,kg}.$

(b) Find the weight of the gold.

$W = mg = (1{,}200\,\text{kg})\left(9.8\,\tfrac{\text{m}}{\text{s}^2}\right) \approx \mathbf{12{,}000\,N}.$

(c) Is it likely that the pirates could carry such a chest?

No! This is over 2600 lbs, about the weight of a small car. Even pirates who have been working out would have a difficult time lifting this.

••Sample Problem 2
A gold prospector finds a solid rock that is composed solely of quartz and gold. The density of quartz is ρ_q and the density of gold is ρ_g.

> Tell me the weight and volume of a piece of quartz and I'll tell you how much gold is in it! Incredible? Yes! Magic? No; just some figuring with physics!

(a) Find the mass of gold contained in the rock of mass M with a volume V.

Focus: $m_{\text{gold}} = ?$
This is a difficult problem to solve because it involves two unknowns—the amount of gold and the amount of quartz. The "trick" is recognizing that the masses of the gold and quartz have to add to M, and that their volumes have to add to V.

    (*Equation* 1) : $m_g + m_q = M$.

    (*Equation* 2) : $V_g + V_q = V$.

What ties mass and volume together is density ρ.

    $\rho_g = \dfrac{m_g}{V_g}$ and $\rho_q = \dfrac{m_q}{V_q}$.

Strategy: We'll write Equation 1 in terms of $m_g = ?$ and let the terms guide us to a solution.

$$m_g = M - m_q$$

$m_g = M - \rho_q V_q$, and we know from Equation 2 that $V_q = V - V_g$. So

$m_g = M - \rho_q(V - V_g)$, and we know that $V_g = \dfrac{m_g}{\rho_g}$. So

$$m_g = M - \rho_q\left(V - \frac{m_g}{\rho_g}\right)$$

$$m_g = M - \rho_q V + \rho_q\frac{m_g}{\rho_g}$$

Rearrange the equation so all of the m_g's are on the left side and we get

$$m_g - \rho_q\frac{m_g}{\rho_g} = M - \rho_q V. \text{ The left side can be expressed } m_g\left(1 - \frac{\rho_q}{\rho_g}\right) = m_g\left(\frac{\rho_g - \rho_q}{\rho_g}\right).$$

Then from $m_g\left(\dfrac{\rho_g - \rho_q}{\rho_g}\right) = M - \rho_q V \;\Rightarrow\; m_g = \dfrac{\rho_g(M - \rho_q V)}{(\rho_g - \rho_q)}.$

How wonderful: The mass of the gold in the rock can be determined by knowing only the mass and volume of the rock, plus the densities of gold and quartz!

(b) Calculate the mass of gold contained in the rock if the rock's mass is 10.00 kg and its volume is 3.50×10^{-3} m³. The density of quartz is 2,650 kg/m³ and the density of gold is 19,300 kg/m³.

Solution:

$$m_g = \frac{\rho_g(M - \rho_q V)}{(\rho_g - \rho_q)} = \frac{19{,}300\,\frac{kg}{m^3}\left[10.00\,kg - 2{,}650\,\frac{kg}{m^3}(3.50\times10^{-3}m^3)\right]}{19{,}300\,\frac{kg}{m^3} - 2{,}650\,\frac{kg}{m^3}} = \textbf{0.840 kg.}$$

This is 8.4% of the total mass of the rock.

(c) How much of the rock's volume is quartz?

Focus: $V_q = ?$

Since we know how much of the rock's mass is gold (0.840 kg) we know that the rest of the rock's mass ($10.00 - 0.84 = 9.16$ kg) must be quartz. Since we know the mass and density of the quartz we can find its volume.

From $\rho = \dfrac{m}{V} \;\Rightarrow\; V_q = \dfrac{m_q}{\rho_q} = \dfrac{9.16\ kg}{2{,}650\,\frac{kg}{m^3}} = 3.46\times10^{-3}m^3.$

Percentage-wise, the quartz makes up

$$\frac{3.46\times10^{-3}m^3}{3.50\times10^{-3}m^3}\times100\% = \textbf{98.9\%} \text{ of the total volume of the rock.}$$

Sample Problem 3

A certain spring having a spring constant k stretches a distance x when it carries a load of mass m.

(a) How much will it stretch when supporting a load $3m$?

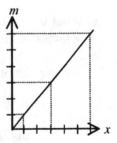

Focus: $x_{3m} = ?$

Three times the mass will be pulled downward by gravity with three times as much force, so we'd expect the stretch to be three times as much.

From $F = kx$ $\Rightarrow x_m = \dfrac{F_m}{k} = \dfrac{mg}{k}$. Likewise, $x_{3m} = \dfrac{F_{3m}}{k} = \dfrac{3mg}{k} = 3\left(\dfrac{mg}{k}\right) = 3x_m$.

(b) How much will it stretch when supporting a load $6m$ (assuming it doesn't reach its elastic limit)?

Focus: $x_{6m} = ?$

By proportional reasoning, twice the load means twice the stretch, or $6x_m$. Or more formally,

$$x_{6m} = \frac{F_{6m}}{k} = \frac{6mg}{k} = 6\left(\frac{mg}{k}\right) = 6x_m.$$

(c) Calculate the answers to (a) and (b) if $k = 29.4$ N/m, $m = 0.040$ kg, and x is 0.012 m.

$x_{3m} = 3x_m = 3(0.012\,\text{m}) = \mathbf{0.036\,m}.$

$x_{6m} = 6x_m = 6(0.012\,\text{m}) = \mathbf{0.076\,m}.$

(d) What would be the shape of a graph of supported mass versus stretch for the spring? Make a rough sketch.

Answer: Because every time m doubles or triples the stretch doubles or triples too, the graph would be a straight line through the origin.

Sample Problem 4

Consider a telephone pole of length L and average diameter D.

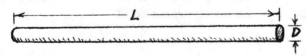

(a) What is its volume?

Answer: The pole is a cylinder, so its volume = (circular area) × length =

$$\pi R^2 L = \pi\left(\tfrac{D}{2}\right)^2 L = \frac{\pi D^2 L}{4}.$$

(b) What is its surface area?

Answer: For a cylinder the surface area is the area of the two circular ends, $2 \times \pi\left(\frac{D}{2}\right)^2$, plus the area of the side. We can imagine that if we were to remove the ends and unroll the side, we'd have a rectangle with a width of πd (the circumference of the circular end) and length L. So the surface area is

$$\frac{\pi D^2}{2} + \pi DL = \pi D\left(\frac{D}{2} + L\right)$$

(c) When scaled up by 2, by how much will its total surface area increase?

Solution: $A_{new} = \pi(2D)\left(\frac{(2D)}{2} + (2L)\right) = 4\pi D\left(\frac{D}{2} + L\right) = 4\,A_{initial}.$

A rule of scaling states that when a body is scaled up by a factor x, the surface area increases by x squared. So scaling up by 2 yields an increase in area of 2^2. The surface area **increases by a factor of 4.**

(d) If it is scaled up by a factor of 2 (twice as long with twice the diameter) by how much will its volume increase?

Solution: $V_{new} = \frac{\pi(2D)^2(2L)}{4} = 8\left(\frac{\pi D^2 L}{4}\right) = 8\,V_{initial}.$

A rule of scaling states that when a body is scaled up by a factor x, the volume increases by x cubed. So the volume of the pole increases by a factor of **8.**

(e) What is the effect on the surface-area-to-volume ratio for the pole when it is scaled up by a factor of 2?

Answer: Since the surface area increases by a factor of 4 and the volume increases by a factor of 8, the surface-area-to-volume ratio, $\dfrac{\text{surface area}}{\text{volume}} = \dfrac{4 \times A_{initial}}{8 \times V_{initial}} = \dfrac{1}{2}\left(\dfrac{A_{initial}}{V_{initial}}\right) = \text{half}$ the original surface-area-to-volume ratio. The same ratio applies to the cross section of the pole compared with its weight. A pole twice as long and twice as wide will have four times the cross-sectional area but eight times the weight, so it will have half the strength compared with its weight. This accounts for the greater sag of the pole when supported horizontally at its ends. What would happen to this ratio if the pole's length and diameter were tripled?

Solids Problems

12-1. A cube of metal with sides of length L is said to be pure platinum. Platinum has a density $2.14 \times 10^4 \ \mathrm{kg/m^3}$.
(a) What is the volume of the platinum cube?
(b) Suppose the cube is 0.35 m on each side. Calculate the mass of the cube.
(c) Calculate the weight of the cube in newtons, and then in pounds (recall that 1 N = 0.22 lb). Could you lift it?

12-2. Linsey has a cube of plastic with sides of length L and mass m.
(a) What is its density?
(b) Calculate its density if its mass is 15 kg and the length of its sides is 0.15 m.

12-3. A block of metal has dimensions l, w, h, and has a mass m.
(a) What is the density of the metal block?
(b) Calculate the density of the block if its dimensions are $l = 0.20$ m, $w = 0.30$ m, $h = 0.15$ m, and its mass is 25 kg.

12-4. A circular slab of ice has a diameter D, thickness d, and density ρ.

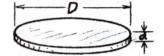

(a) What is the mass of the ice?
(b) Calculate the mass if the diameter of the slab is 5.00 m, its thickness is 0.020 m, and its density is 917 $\mathrm{kg/m^3}$.

12-5. A liter of a certain liquid has a mass m.
(a) What is its density?
(b) Calculate its density if its mass is 13.6 kg. (Recall that $1 \ \mathrm{L} = 10^{-3} \ \mathrm{m}$.)

12-6. A sphere of a certain material has a mass m and radius r.
(a) What is its density in units $\mathrm{kg/m^3}$? (The volume of a sphere is $4/3 \ \pi r^3$.)
(b) Calculate the sphere's density if its mass is 35 kg and its radius is 0.22 m.

12-7. Neutron stars have incredibly high densities.
(a) Calculate the density of a spherical neutron star of mass 3.0×10^{28} kg and radius 1.3×10^3 m.
(b) What is the mass of one teaspoon of this neutron star? One teaspoon is about 5 cubic centimeters.

12-8. A uniform lead sphere and a uniform aluminum sphere have the same volumes. The density of lead is 11,344 $\mathrm{kg/m^3}$ and the density of aluminum is 2,700 $\mathrm{kg/m^3}$.
(a) What is the ratio of densities for the lead and aluminum spheres?
(b) What is the ratio of masses?

12-9. A uniform lead sphere and a uniform aluminum sphere have the same mass.
(a) What is the ratio of radii for the lead and aluminum spheres?
(b) What is the ratio of their diameters?

12-10. • A silver prospector finds a solid rock that is composed solely of quartz and silver. The density of quartz is 2,650 kg/m³ and the density of silver is 10,500 kg/m³.
  (a) Calculate the mass of silver contained in the rock of mass 5.50 kg with a volume of 2.00×10^{-3} m³.
  (b) How much of the rock volume is quartz?

12-11. Silver can be pounded into extremely thin sheets.
  (a) If the silver can be pounded into a sheet of thickness t, what is the area A of the sheet?
  (b) Calculate the area of such a sheet 3.00×10^{-7} m thick made from 0.500 kg of silver.

12-12. The density of water is 1000 kg/m³. The density of ice is 917 kg/m³.
  (a) When a given mass of water changes phase to form ice, by what factor does its volume increase?
  (b) Which weighs more, a liter of water or a liter of ice (recall 1 L = 1000 cm³)?

12-13. A spring with a spring constant k is compressed a distance x when a load of mass m is placed on it.
  (a) How much will the spring compress when supporting a load $4m$?
  (b) How much will it compress when supporting a load $8m$ (assuming it doesn't reach its elastic limit)?
  (c) Calculate the answers to (a) and (b) if $k = 392$ N/m, $m = 2.00$ kg, and x is 0.050 m.

12-14. A truck of mass M hauls a trailer of mass m connected by a spring. The spring has a spring constant k. The truck accelerates from rest at a constant rate to speed v in time t while the stretched spring maintains a constant length.
  (a) What is the acceleration of the truck and trailer?
  (b) By how much is the spring stretched?
  (c) Does the mass of the truck matter? Defend your answer.
  (d) Calculate answers to (a) and (b) above for a 3200 kg truck pulling a 420 kg trailer from rest to 22 m/s in 15.0 seconds using a spring with $k = 3900$ N/m.

12-15. An object of mass m is suspended from a spring of spring constant k that hangs from the ceiling of an elevator. How much does the spring stretch:
  (a) When the elevator is at rest?
  (b) When the elevator moves upward at a constant speed v?
  (c) When the elevator accelerates upward at a?
  (d) When the elevator accelerates downward at a?
  (e) When the elevator moves downward at a constant speed v?
  (f) Calculate the spring stretch for the case of a 4.5-kg object suspended from a spring having a spring constant $k = 750$ N/m in the elevator at rest.
  (g) Calculate the spring stretch for the same mass when the elevator accelerates upward at 0.50 m/s².
  (h) Calculate the spring stretch for the same mass when the elevator accelerates downward at 0.50 m/s².

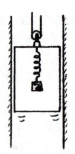

12-16. A block of mass m is pulled across a frictionless horizontal surface by a spring having a spring constant k. The block undergoes an acceleration a.
(a) What is the magnitude of the net force on the block?
(b) By how much does the spring stretch?
(c) Calculate the net force on the block and the stretch of the spring for a 1.5-kg block being pulled by a spring with $k = 150$ N/m with an acceleration of 0.30 m/s^2.

12-17. A ball of mass m is whirled in a horizontal circle by a rope attached to a spring having a spring constant k. The length of rope including the spring is L and the speed of the ball is v.

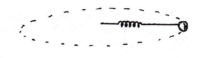

(a) What is the magnitude of the centripetal force, assuming the string is practically horizontal?
(b) By how much is the spring stretched?
(c) Calculate the magnitude of the centripetal force and the stretch in the spring with $k = 720$ N/m for a 0.36 kg ball being whirled at 7.0 m/s in a horizontal circle of radius 0.49 m.

12-18. A block of mass m rests on a horizontal air table (friction free) and is attached to a practically massless horizontal spring. When the block is pulled horizontally from rest by the spring with the spring maintaining a constant stretch, the block reaches a speed v in time t.
(a) How much force acts on the block?
(b) By how much does the spring stretch?
(c) Calculate the force on a 4.52 kg block and the stretch in the spring with $k = 120$ N/m if the block reaches a speed of 0.45 m/s from rest in 1.3 seconds.

12-19. A spring of length L has a spring constant k. The spring is cut in half. A load of mass m is hung by one of the half-length vertical springs.
(a) What is the tension in the spring of length L when supporting the load m?
(b) What is the tension in a half-length spring supporting the same load? (Hint: It's okay that these questions are easy ones with easy answers.)
(c) How far will the half-length spring stretch compared with the spring before being cut in half?
(d) What is the spring constant of each spring of length $L/2$?

12-20. Consider a rectangular beam of length L and cross-sectional area A.

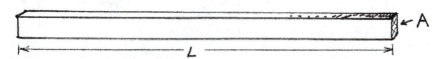

(a) What is its volume?
(b) If its linear dimensions are scaled up by a factor of 3 how will its volume increase?
(c) When scaled up by 3, by how much will its total surface area increase?
(d) What is the effect of scaling up the linear dimensions by 3 on the surface-area-to-volume ratio for the beam?
(e) Will the scaled-up beam be more or less likely to sag when supported at its ends?

12-21. A plastic cube has sides of length L.
    (a) What is the area of one of its sides?
    (b) What is the total surface area of the cube?
    (c) What is the volume of the cube?
    (d) What is the ratio of total surface area to volume?
    (e) For a larger cube, is the total-surface-area-to-volume ratio more or less?

12-22. A cube of cheese with sides L is sliced into cubes with sides $L/2$.
    (a) How many cubes result?
    (b) What is the area of each side of the smaller cube?
    (c) What is the total surface area of each smaller cube?
    (d) What is the sum of total surface areas of the smaller cubes, and how does it compare
        with the total surface area of the single cube with sides L?
    (e) What is the volume of each smaller cube?
    (f) What is the surface-area-to-volume ratio of the cube with side L?
    (g) What is the surface-area-to-volume ratio for each of the cubes with side $L/2$?
    (h) How does the surface-area-to-volume ratio change as the cube is cut into smaller and
        smaller cubes?

12-23. A spherical blob of mercury breaks into two identical smaller spherical blobs. (The
    surface area of a sphere is $4\pi r^2$, and the volume $\frac{4}{3}\pi r^3$.)
    (a) How does the volume of the larger blob compare with that of each smaller blob?
    (b) How does the surface area of the larger blob compare with that of each smaller blob?
    (c) What is the surface-area-to-volume ratio for the larger blob?
    (d) What is the surface-area-to-volume ratio for each of the smaller blobs?
    (e) If each of the blobs is forced into a cube shape, how is the surface-area-to-volume
        ratio affected?
    (f) If the blobs break into even smaller ones, how is the surface-area-to-volume ratio
        affected?

12-24. Four little spheres of mercury, each with a radius r, coalesce to form a single sphere.
    (a) What will be the radius of the coalesced sphere?
    (b) How does the surface area of the coalesced sphere compare with the total surface
        areas of the four smaller ones?
    (c) Has the area/volume increased or decreased for the coalesced sphere?

12-25. A storage tank is doing a good job supplying water to a population
    of people. The design of a new storage tank is being considered in
    a location that serves half the population. A rookie designer
    proposes that the new tank have half the linear dimensions of the
    present tank.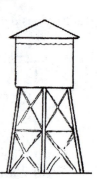
    (a) How much smaller is the volume of the new tank compared
        with that of the present tank?
    (b) For painting purposes, how much smaller is the overall surface
        area of the new tank?
    (c) Why was the rookie's proposal rejected?

Show-That Problems

12-26. A 2.900-kg solid cylinder that is 0.1200 m tall has a radius of 0.0300 m.
Show that the density of the cylinder is about 8550 kg/m^3.

12-27. Cork has a density of 300 kg/m^3.
Show that a cubic meter of cork would be too heavy for a person to lift.

12-28. A solid iron sphere ($\rho = 7{,}874$ kg/m^3) has a diameter of 0.040 m.
Show that its mass is 0.26 kg.

12-29. A queen orders a 0.2000-kg crown made from pure gold. When it arrives from the
crown maker the volume of the crown is found to be 1.0363×10^{-5} m^3. (Recall that
$\rho_g = 19{,}300$ kg/m^3.)
Show that the crown is indeed gold.

12-30. A solid copper sphere of radius 0.0150 m is suspended from a vertical rubber band. The
density of copper is 8,960 kg/m^3.
Show that the tension in the rubber band is 1.24 N.

12-31. A full half-liter can of water has a mass of 0.525 kg.
Show that the volume of aluminum to make the can is 9.26×10^{-6} m^3. (Recall that the
density of aluminum is 2,700 kg/m^3.)

12-32. Note data for Earth on the inside back cover of your textbook.
Show that the average density of Earth is approximately 5,500 kg/m^3.

12-33. Note data for the Sun on the inside back cover of your textbook.
Show that the average density of the Sun is approximately one-fourth that of Earth's
density.

12-34. Some neutron stars have the same mass as the Sun but with much smaller radii.
Show that the density of a neutron star having the mass of the Sun but a radius of
20.0 km is 5.94×10^{16} kg/m^3.

12-35. A crystal cube has a mass of 12.0 kg.
Show that the same cube scaled down to half size would have a mass of 1.5 kg.

12-36. A bust of a famous president cast in solid brass has a mass of 12.0 kg.
Show that the same bust scaled down to half size would have a mass of 1.5 kg (using the
logic of the previous problem).

12-37. A model steel bridge is 1/20 the exact scale of the real bridge that is to be built.
Show that if the model bridge weighs 50 N that the real bridge will weigh about 400,000 N.

12-38. Toothpicks are made from a log that is 100 times longer and 100 times thicker than an
individual toothpick.
Show that the weight of the log is a million times greater than the weight of one of the
toothpicks.

12-39. When supported horizontally at both ends, the toothpick of the previous problem shows no noticeable sag. But the log from which it was made shows a noticeable sag when similarly supported.

Show that each unit of cross-sectional area for the log supports 100 times more weight than each unit of cross section for the toothpick.

12-40. Two people at the beach, one twice as heavy as the other, apply suntan lotion to their bodies.

Show that the amount of lotion needed by the twice-as-heavy sunbather is about 1.6 times as much.

13 Liquids

The physics of liquids is perhaps the oldest body of physics knowledge. And it is often overlooked in physics courses. The chapter begins with pressure, defined as the force per unit area. The SI unit for pressure is the *pascal* (Pa). 1 Pa = 1 N/m². One pascal is a small pressure. Standard atmospheric pressure is 1.01×10^5 Pa. The key player in this chapter is Archimedes' principle, which states "An immersed body is buoyed up by a force equal to the weight of the fluid it displaces." We call this upward force a *buoyant force*. If whatever you place into a fluid displaces, say, 2 N of fluid, the object will experience a upward force of 2 N exerted on it by the fluid. Understanding of Archimedes' principle requires a clear distinction between weight and volume. We speak of a body displacing its *volume* of fluid when submerged in a liquid, and the buoyant force being equal to the *weight* of the volume of liquid displaced. Although the focus of this chapter is liquids, Archimedes' principle applies to gases as well, as we'll observe in the following chapter.

We end the chapter with Pascal's principle, which states: "A change in pressure at any point in an enclosed fluid at rest is transmitted undiminished to all points in the fluid." We use the Greek letter ρ ("rho") for density. The density of water is 1000 kg/m³.

Sample Problem 1

A swimming pool has a depth h and cross-sectional area A.

(a) What is the pressure due to water at the bottom of the pool?

Focus: $p = ?$

Taking only the water into account, the pressure at the bottom of the pool is due to the weight of the water spread over the area of the bottom of the pool.

$$p = \frac{m_w g}{A} = \frac{(\rho_w V_w)g}{A} = \frac{\rho_w(Ah)g}{A} = \rho_w gh.$$

Notice that the pressure is independent of the actual area of the pool. It only depends upon the depth of the water and its density. Let's confirm that ρgh yields units of pressure:

$$\rho\left(\tfrac{kg}{m^3}\right) g\left(\tfrac{m}{s^2}\right) h(m) = \frac{\left(kg \cdot \tfrac{m}{s^2}\right) \cdot m}{m^3} = \frac{N}{m^2} = Pa.$$

(b) What is the total pressure at the bottom of the pool if atmospheric pressure is P?

Focus: $p = ?$

The total pressure at the bottom of the pool will be the pressure due to the atmosphere plus the pressure due to the water.

$$p = P + p_w = P + \rho_w gh.$$

(c) Calculate the pressure at the bottom of the pool if it's 2.5 m deep, has a cross-sectional area of 50.0 m², and atmospheric pressure is 1.01×10^5 Pa.

$$p = P + \rho_w gh = 1.01 \times 10^5 Pa + \left(1000 \tfrac{kg}{m^3}\right)\left(9.8 \tfrac{m}{s^2}\right)(2.5 m) = \mathbf{1.26 \times 10^5 \, Pa}.$$

(This is 1.24 atmospheres of pressure.)

Sample Problem 2

A chunk of ore is suspended by a light string has weight W in air. When the chunk is totally immersed in a container of water, tension in the string is w.

(a) What is the buoyant force acting on the immersed chunk of ore?

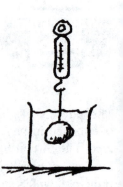

Analysis: Here we refer back to Newton's first law—since the chunk is at rest the net force acting on it must be zero.

The upward forces on the ore are the tension in the string and the buoyant force; the downward force is the ore's weight.

$$\Sigma F = \text{Tension} + BF - \text{Weight} = 0 \quad \Rightarrow w + BF - W = 0 \quad \Rightarrow BF = W - w.$$

(b) Suppose the chunk is then totally immersed in another liquid of unknown density, and the tension in the string is $0.9w$. What is the density of the unknown liquid?

Analysis: Since the tension in the string is less, the new liquid must be exerting a greater buoyant force for the same *volume* of liquid displaced, so the new liquid must have a higher density (for example, if the liquid were twice as dense, the weight of the liquid displaced by the ore would be twice as much, so the buoyant force would be twice as much).

$$\frac{BF_{new}}{BF_0} = \frac{W - 0.9w}{W - w} \text{ and } \frac{BF_{new}}{BF_0} = \frac{\rho_{new}}{\rho_0} \quad \Rightarrow \frac{\rho_{new}}{\rho_0} = \frac{W - 0.9w}{W - w} \quad \Rightarrow \rho_{new} = \rho_0 \left(\frac{W - 0.9w}{W - w} \right).$$

Sample Problem 3

A pie pan with negligible mass floats on the surface of water in a small aquarium tank. When a piece of iron of density ρ and mass m is placed in the pie pan (which remains floating) the water level rises.

(a) What is the weight of water that rises?

Answer: The pan floats at rest, so the buoyant force on the pan is equal to the weight of the piece of iron. By Archimedes' principle, the weight of the water displaced = the weight of the piece of iron = $m_{iron}g$.

(b) What is the volume of water that rises?

Focus: $V_w = ?$ From $\rho = \dfrac{m}{V} \Rightarrow V_w = \dfrac{m_w}{\rho_w}$. From above, the weight of the water displaced is equal to the weight of the iron. So the mass of the water displaced is equal to the mass of the iron: $V_w = \dfrac{m_{iron}}{\rho_w}$.

(c) What volume of water would rise if the iron were not floating, but at the bottom of the tank?

Focus: $V_w = ?$ This time the water displaced has the same *volume* as the iron, not the same *weight*. The volume of the iron, and therefore of the water displaced, is $V_w = \dfrac{m_{iron}}{\rho_{iron}}$. Since iron is almost 8 times more dense than water, the volume of water displaced when the iron is submerged is about one-eighth the volume displaced when it is floating in the pie pan.

(d) When the iron "falls overboard" from the pie pan and sinks in the tank, does the water level of the tank rise, fall, or remain unchanged?

Answer: The water level **falls** compared with the water level when the iron floats in the pan. As mentioned in (c), less water is displaced when the iron is submerged than when it is in the floating pan.

Sample Problem 4
In order to lift a heavy load, a hydraulic machine injects oil into a closed cylinder with a pressure $p + P$, where P is atmospheric pressure and p is the so-called "gauge pressure," the pressure beyond atmospheric. The plunger in the cylinder has a radius r, and has negligible mass compared with the loads to be lifted.
(a) What force is exerted on the plunger?

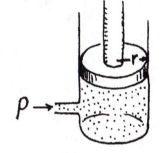

Focus: $F = ?$ The upward pressure on the plunger from the oil below it is $p + P$ and the downward pressure from the air above it is P, so the net upward pressure is p, which results in an upward force on the plunger. (We assume that changes of pressure within the closed cylinder caused by gravity can be neglected relative to the high pressure $p + P$. In other words, we take the pressure $p + P$ to be uniform within the cylinder.)

From *Pressure* $= \dfrac{F}{A}$ $\Rightarrow F = pA = p\pi r^2$.

(b) Calculate the upward force exerted on the plunger if it has a radius 0.120 m and the gauge pressure input is 3.25×10^6 N/m².

$$F = p\pi r^2 = \left(3.25 \times 10^6 \,\tfrac{N}{m^2}\right)\pi(0.12\,m)^2 = 1.5 \times 10^5 \, N.$$

(c) How many kilograms of mass can be lifted by this machine?

Focus: $m = ?$ If the machine is lifting at its maximum capacity, the lifting force of the plunger will be exactly balanced by the weight of the mass being lifted.

$$F_{machine} = mg \Rightarrow m = \frac{F_{machine}}{g} = \frac{1.5 \times 10^5 \, N}{9.8\,\tfrac{m}{s^2}} = 1.5 \times 10^4 \, kg.$$

(d) Would the machine be able to lift a cube of granite, 1.7 m on a side? Granite has a density 2750 kg/m³.

Answer: Yes. The mass of such a cube $= \rho V = (2750 \, kg/m^3)(1.7\,m)^3 = 14,00$ kg, which is less than 1.5×10^4 kg.

Problems

13-1. When you stand with two feet on a bathroom scale, your weight is given by the pointer or digital readout on the scale.
(a) What happens to the pressure on the scale when you stand on one foot?
(b) Does the scale show this increase in pressure? Defend your answer.

13-2. While lying on a lawn, a petite friend comes by and wishes to test the strength of a boyfriend's stomach muscles. She stands on his stomach.
(a) Calculate the pressure she exerts if her weight is 100.0 N and the area of her two feet is 0.020 m^2.
(b) Calculate the pressure she exerts if she's wearing spike heels, and stands on only the heels of combined area 0.0002 m^3.
(c) Why are spike heels not a good idea on a new linoleum floor (or on your stomach)?

13-3. The circular glass plate on a scuba diver's mask has a radius r. Above the surface of a lake, the only force on the outside of the glass plate is that due to the pressure of the atmosphere. (Atmospheric pressure is normally 1.01×10^5 Pa.)
(a) What is the force on the outside of the glass plate due to atmospheric pressure? Explain why this is not a net force.
(b) The diver then dives to a depth D in the lake. What is the force on the mask due only to water?
(c) Calculate answers to parts (a) and (b) if the radius of the mask is 0.070 m, and water depth D is 10.0 m.

13-4. A slab of ice, 0.92 times the density of water, floats on a fresh-water lake.
(a) What minimum volume must the slab have for a 30-kg youngster to be able to stand on it without getting her feet wet?
(b) If the slab is 0.050 m thick, what must be the cross-sectional area?

13-5. A coal barge of mass m floats in fresh water. When empty, the waterline is shown by the dashed line in the sketch.

When loaded, the line is beneath the water surface. The area of the bottom of the barge is A, and the mass of coal is M.
(a) What is the buoyant force keeping the loaded barge afloat?
(b) What is the volume of water displaced?
(c) What is the added depth of the barge when loaded with coal?
(d) Give numerical values to the above given that the empty barge has a mass of 5000 kg, the area of the bottom of the barge is 30 m^2, and the mass of the coal is 10,000 kg.

13-6. A rock suspended by a spring scale in air weighs W. When suspended while submerged in water it weighs w.*
(a) Find the volume of the rock.
(b) Find the density of the rock.
(c) Calculate these values if the rock weighs 16.5 N in air and 10.3 N in water.

---

* "Weighs in water" means "This is what the scale reads when the rock is submerged in the water."

13-7. A metal cylinder of weight w is suspended in a container of water of weight W, fully immersed without touching the bottom. The container is placed on a weighing scale.

(a) Will the reading on the scale be $w + W$, less than $w + W$, or more than $w + W$?

(b) What will be the reading if the cylinder is released and allowed to rest on the bottom of the container of water?

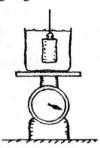

13-8. In lab you find that a piece of metal has weight W on a triple-beam balance, but when suspended in water from a spring balance it has weight w.

(a) Find the density of the metal.

(b) Calculate the density if the metal "weighs" 0.060 kg in air, but only 0.050 kg when suspended in water.

(c) Why are quotation marks around the word *weighs* in (b)?

13-9. A wooden block of mass m is placed on a scale next to a beaker of water, which has a mass M.

(a) What is the reading on the scale in newtons?

(b) If the block is placed in the beaker and floats, what will be the reading on the scale?

(c) If the block is instead a solid cube of iron of mass $8m$, what will be the reading on the scale when the cube is beside the beaker?

(d) What will be the reading if the iron cube is sitting on the bottom of beaker?

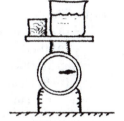

13-10. A wooden ball of mass m and volume V is held beneath the surface of water in a swimming pool.

(a) What is the buoyant force acting on the ball?

(b) How much force must be exerted to hold the ball down?

(c) What would be the tension in a cord attached to the bottom of the pool holding the ball motionless beneath the surface?

(d) Are your answers to (b) and (c) the same or different? Why?

(e) Calculate the tension in the cord when a wooden ball of mass 0.056 kg and volume 7.2×10^{-5} m^3 is attached to the bottom of a water-filled pool.

13-11. A lump of clay has twice the density of water. When it is put in water it sinks.

(a) When the clay is sunk, is the volume of water displaced less than, the same as, or greater than the clay's volume?

(b) The clay is removed and reformed into the shape of a bowl or a boat—which floats on water. Is the volume of water displaced by the clay less than, the same as, or greater than the volume of clay?

(c) Calculate the buoyant force on the clay of mass 1.0 kg when submerged in water.

(d) Calculate the buoyant force on it when it is shaped like a boat and floats.

13-12. A fish of weight w is in the middle of a container of water of weight W. The container with the fish in it is placed on a weighing scale.

    (a) What is the average density of the fish? Defend your answer.
    (b) What is the volume of the fish?
    (c) What is the reading on the weighing scale?
    (d) What will be the scale reading if the fish expands its air sac and floats on the water?

13-13. A U-tube is fitted with pistons to make a simple hydraulic machine. The area of the small piston is A and the area of the larger output piston is $20\ A$. A pressure P is exerted on the smaller piston.
    (a) What is the pressure exerted on the larger piston?
    (b) If force F acts on the smaller piston to produce pressure P, how much force acts on the larger piston?
    (c) Calculate the force acting on the larger piston if a 450-N force acts on the smaller piston, whose area is 12 cm^2.

13-14. The output piston of a hydraulic press has a cross-sectional area A.
    (a) How much fluid pressure is needed to produce a force F on the output piston?
    (b) If the area of the input piston is 0.20A, what force is needed on the input piston to produce a force F on the output piston?
    (c) Calculate the input force needed to produce an output force of 1.5×10^6 N for a 0.25 m^2 output piston.

13-15. In an automotive service station a car is hoisted with a hydraulic press. Pressurized fluid is pumped into a cylinder below a piston that is raised, and which raises the car.

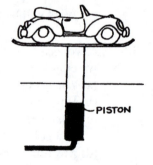

    (a) If the fluid is pressurized to a gauge pressure P, and the area of the piston that raises the car is A, what weight can the piston support?
    (b) Calculate the weight the piston can support if input gauge pressure is 3.5×10^6 Pa, the area of the piston is 0.40 m^2, and the weight of the hydraulic fluid is neglected (recall that 1 Pa = 1 N/m^2).

13-16. A backhoe uses a hydraulic cylinder to lift heavy loads. A pump injects hydraulic oil into a cylinder at a gauge pressure P and drives an output plunger, which has a cross-section A.
    (a) What force can the output plunger exert?
    (b) Calculate this force if the input gauge pressure is 3.3×10^6 Pa and the cross-section of the piston is 0.070 m^2.

Show-That Problems

13-17. An air mattress floating in a pool has dimensions 2.0 m × 0.60 m × 0.20 m, and has a mass of 1.0 kg (including the air within it).
Show that it can support (barely) a group of children with a total mass of 230 kg but that if another child with a mass of 30 kg joins the party, it will sink.

13-18. A flat-bottomed swimming pool has an area of 30.0 m × 10.0 m and a depth of 2.0 m.
Show that the total weight of water in the pool when full is about 6.0×10^6 N, and that the pressure due to this weight of water on the pool bottom is nearly 20,000 Pa (recall that 1 Pa = 1 N/m^2).

13-19. A water tower supplies water to The Villages in Florida. The tower is 50 m high.
Show that the pressure at the base of the tower due to the water is nearly 500,000 Pa.

13-20. Water next to the Hoover Dam is 221 m deep.
Show that the pressure at the base of the dam due to the water is about 2.2 million Pa.

13-21. The vertical face of a reservoir dam is 100.0 m wide and 12.0 m high.
Show that when water reaches the crest of the dam the average water pressure on the face is about 60,000 Pa.

13-22. The Mariana trench off the coast of the Philippines is about 11,000 m deep. The average density of seawater is 1025 kg/m^3.
Show that the pressure on the window of a submersible at that depth is more than 1000 times atmospheric pressure.

13-23. The density of seawater is 1.03×10^3 kg/m^3. The density of ice is 0.92×10^3 kg/m^3.
Show that 11% of the total volume of an iceberg extends above water level.

13-24. An aluminum cube 0.15 m on each side is submerged in water.
Show that the buoyant force that acts on the cube is 33 N.

13-25. A sample of pure gold of mass 1.20 kg is placed in a container of water.
Show that the volume of water displaced is 6.22×10^{-5} m^3. (The density of gold is 19.3 times that of water.)

13-26. A log floats in water with one-fourth of its volume above the surface.
Show that the density of the log is 750 kg/m^3.

13-27. A ferryboat is 4.0 m wide and 6.0 m long. When Fat Freddy steps on board the boat sinks 0.0060 m deeper in the water.
Show that Fat Freddy weighs about 1400 N.

13-28. A rectangular air mattress of dimensions 2.0 m × 0.60 m × 0.10 m floats in a swimming pool.
Show that, including its own mass, the maximum mass it can support before being completely submerged is 120 kg.

13-29. A novice weightlifter interested in aerobics can just barely lift a 30-kg iron barbell over his head in the gym.

Show that under water the weightlifter can lift an iron barbell of nearly 34 kg. (The density of iron is 7,874 kg/m^3.)

13-30. A submarine has a weight of 10,000 tons. When floating partially submerged it displaces 10,000 tons of water.

Show that, using reasoning, appreciably more than 10,000 tons of water must be displaced when in equilibrium beneath the surface.

13-31. • A 1967 Kennedy half-dollar is a mixture of silver and copper and has a mass of 0.01150 kg. In water the coin weighs 0.1011 N.

Show that the coin contains nearly equal volumes of silver and copper. (The density of silver is 10,490 kg/m^3; the density of copper is 8,920 kg/m^3.)

13-32. A U-tube with pistons fitted at both ends comprises a hydraulic machine.

Show that the ratio of input to output piston areas A_1/A_2 is equal to the ratio of input to output forces F_1/F_2.

13-33. A force of 100 N is applied to the small cylinder of the hydraulic device shown. The output cylinder has a diameter 2.5 times greater than the input cylinder.

Show that the output force is 625 N.

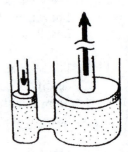

13-34. The piston in the small cylinder of the previous problem moves 6.25 cm downward when it does work on the system.

Show that the output cylinder moves a distance of 1.0 cm.

13-35. A machine uses a hydraulic cylinder to lift a heavy load. Hydraulic oil is injected into the cylinder at a gauge pressure of 4.06×10^6 Pa (which means over and above atmospheric pressure), which drives the output plunger. The output plunger has a radius of 0.120 m.

Show that the weight the device can lift is 1.84×10^5 N.

14   *Gases and Plasmas*

Much of the physics of liquids applies to gases, for both are fluids—gases being much less dense than liquids. A big difference between liquids and gases is that in many applications, the density of a liquid remains essentially constant over a huge range of pressures, whereas the density of a gas is quite sensitive to changes in pressure. At constant temperature, gas density and pressure are related by *Boyle's law*, $P_1/\rho_1 = P_2/\rho_2$, which can also be written $P_1V_1 = P_2V_2$, where P is pressure, ρ is density, and V is volume.

The physics of fluids in motion is described by Bernoulli's principle, $P + \frac{1}{2}\rho v^2 = $ constant in steady horizontal flow. This means that for water flow in a horizontal pipe, $P + \frac{1}{2}\rho v^2$ in one section of the pipe is equal to $P + \frac{1}{2}\rho v^2$ in another. So as speed increases, pressure decreases (and vice versa).

Sample Problem 1

Air has weight. The density of 20°C air at sea level is 1.20 kg/m³.

(a) Calculate the weight of air in a typical school gym with rectangular dimensions 60.0 m × 30.0 m × 10.0 m.

Focus: $W = ?$ From the dimensions of the room we can figure out the volume of the air, and from the volume and density we can determine the mass.

From $\rho = \dfrac{m}{V} \Rightarrow m = \rho V.$ $W = mg = \rho V g = \rho(l \times w \times h)g = 1.20\frac{kg}{m^3}(60.0\,m \times 30.0\,m \times 10.0\,m)\left(9.8\frac{m}{s^2}\right)$

$= 212,000\,N.$

(b) What would this weight of air be if it were compressed?

Answer: Compression of air changes volume but not weight. It would still weigh **212,000 N**.

(c) Could you lift this weight of air if it were compressed?

Answer: **No**. This weight of air is over 20 tons!

Sample Problem 2

A buoyant force acts on us at all times. It also acts on inflated balloons.

(a) Lillian has a mass of 47 kg. What approximate volume of air of density 1.20 kg/m³ matches her mass?

From $\rho = \dfrac{m}{V} \Rightarrow V = \dfrac{m}{\rho} = \dfrac{47\,kg}{1.20\frac{kg}{m^3}} = 39\,m^3.$

(b) How big would a helium-filled balloon need to be to lift Lillian? Neglect the buoyancy on Lillian herself, assume the mass of the balloon itself is negligible, and that the density of helium is 0.178 kg/m³.

Focus: $V = ?$ If the gas inside the balloon had no mass, a 39 m³ balloon would suffice to displace a weight of air equal to Lillian's weight. But since the helium inside has weight, the balloon has to displace a larger volume of air to lift this added weight.

From $\rho = \dfrac{m}{V} \Rightarrow V_{\text{balloon}} = \dfrac{m_{\text{air displaced}}}{\rho_{\text{air}}}$.

We know ρ_{air}. What is $m_{\text{air displaced}}$?

From $BF = m_{\text{air displaced}} g = m_{\text{Lillian}} g + m_{\text{Helium}} g \Rightarrow m_{\text{air displaced}} = m_{\text{Lillian}} + m_{\text{helium}}$.

We know m_{Lillian}.

Now we need m_{helium}. From $\rho = \dfrac{m}{V} \Rightarrow m_{\text{helium}} = \rho_{\text{helium}} V_{\text{balloon}}$.

Putting all of this together,

$$V_{\text{balloon}} = \frac{m_{\text{air displaced}}}{\rho_{\text{air}}} = \frac{m_{\text{Lillian}} + \rho_{\text{helium}} V_{\text{balloon}}}{\rho_{\text{air}}} \Rightarrow \rho_{\text{air}} V_{\text{balloon}} - \rho_{\text{helium}} V_{\text{balloon}} = m_{\text{Lillian}}.$$

$$\Rightarrow V_{\text{balloon}} = \frac{m_{\text{Lillian}}}{\rho_{\text{air}} - \rho_{\text{helium}}} = \frac{47\ \text{kg}}{1.20\,\frac{\text{kg}}{\text{m}^3} - 0.178\,\frac{\text{kg}}{\text{m}^3}} = \mathbf{46\ m^3}.$$

This is a balloon with a diameter of just about 4.5 meters.

Sample Problem 3
Water flows through a horizontal pipe into a horizontal narrower pipe. Water pressure is P where the speed is v_1 through the larger cross-sectional area A_1. The flow continues through the smaller section of cross-sectional area A_2.
(a) Find the speed where the area of the pipe A_2 is one-fourth the area of A_1.

Analysis: Let's consider a "chunk" of fluid flowing through the pipe. We see the chunk at some time t_1 where the pipe has a cross-sectional area A_1, and the same chunk a little while later at time t_2 where the pipe has a cross-sectional area A_2. If the fluid is *incompressible* (as with water, where its density remains constant even as the pressure changes) then the volume of the fluid at these two times must be the same. So, referring to the diagram, $A_1 \Delta x_1 = A_2 \Delta x_2$.

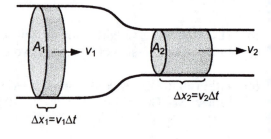

It takes the same amount of time for the fluid to travel a distance Δx_1 in the wider section of pipe as it does for that same volume of fluid to travel a distance Δx_2 in the narrower section of pipe. So $A_1 \Delta x_1 = A_2 \Delta x_2$ becomes $A_1(v_1 \Delta t) = A_2(v_2 \Delta t) \Rightarrow A_1 v_1 = A_2 v_2$. This last expression can be thought of as a restatement of the law of conservation of mass for an incompressible moving fluid, and is called the *equation of continuity*.

For our particular problem we have $A_2 = A_1/4$, and we're looking for v_2.

From $A_1 v_1 = A_2 v_2 \Rightarrow v_2 = \dfrac{A_1}{A_2} v_1 = \dfrac{A}{\left(\frac{A}{4}\right)} v_1 = \mathbf{4v_1}.$

(b) If the pressure in the wide part of the pipe is P_1, what is the pressure in the narrower section of pipe?

Analysis: Bernoulli's principle relates pressures changes to speed changes. Since in steady horizontal flow,

$P_1 + \frac{1}{2}\rho v_1^2 = P_2 + \frac{1}{2}\rho v_2^2$, it follows that the pressure in the narrower pipe section is $P_2 = P_1 - \frac{1}{2}\rho(v_2^2 - v_1^2)$. In this example $v_2 = 4v_1$, so $P_2 = P_1 - 7.5\ \rho v_1^2$.

(c) If the speed of the water in the wider pipe is 4 m/s, what is the change of pressure from the wider to the narrower pipe? How many atmospheres of pressure change is this?

Solution: $P_2 - P_1 = -7.5\rho v_1^2 = -7.5\left(1000\frac{kg}{m^3}\right)\left(4\frac{m}{s}\right)^2 = $ **−120,000 Pa**.

Divide this by atmospheric pressure, 101,000 Pa, to get a pressure **decrease of about 1.2 atmospheres**.

Problems

14-1. The dimensions of a typical living room are 4.0 m ×5.0 m ×2.5 m.
    (a) What is the weight of air in the room?
    (b) What are the mass and weight of an equal volume of water?
    (c) Would a typical floor be able to support this weight of water? Defend your answer, with the knowledge that floors are typically designed to support a static load of 1900 N/m² or so.

14-2. A Cartesian diver consists of a water-filled plastic bottle with a partially filled eyedropper submerged in it. When you squeeze the bottle, the pressure inside increases, which compresses the air inside the dropper. The water moving into the dropper makes the dropper denser—and it sinks. When you stop squeezing the bottle a decrease in pressure results, density of the dropper decreases, and the eyedropper moves upward.

    (a) When the eyedropper is motionless, how does buoyant force on it compare with its weight (including the water in it)?
    (b) Can the eyedropper remain motionless with more water inside it? Defend your answer.
    (c) Calculate the buoyant force on the dropper when it is motionless if its mass with water inside is m.

14-3. An air-filled party balloon is squeezed to one-third its volume (without changing its temperature).
    (a) Compared with the pressure before squeezing, what is the new pressure of air in the balloon?
    (b) Compared with its mass before squeezing, what is the mass of air after being squeezed?
    (c) Compared with the density before squeezing, what is the density of air in the balloon after being squeezed?

14-4. Air in a cylinder is compressed to one-fifth its original volume with no change in temperature.
(a) Find the pressure compared with the original pressure.
(b) Find the density of the compressed gas compared with the original density.
(c) Calculate the new pressure and density if the original pressure was 1.1×10^5 Pa and the original density was 1.3 kg/m^3.

14-5. Air pressure is nicely demonstrated with Magdeburg hemispheres. Atmospheric pressure is P and the cross-sectional area of the circle where the hemispheres meet is A. (Interestingly, the net force over the curved surface of a hemisphere is the same as over a flat disk of area A.)

(a) When the hemispheres are fully evacuated, what force is needed to pull them apart?
(b) The air between the hemispheres is initially at atmospheric pressure. What force is needed to separate them when 80% of the air molecules have been evacuated from the hemispheres?
(c) Calculate these values when atmospheric pressure is 1.01×10^5 Pa and the cross-sectional area A is 0.200 m^2.

14-6. Professor John tosses a rubber mat onto the top of a stool. With a hook secured to the top of the mat, he lifts the mat and stool together.

(a) What is your explanation for the lifting of the stool?
(b) If the area of the rubber mat is 0.90 m^2, what is the maximum weight it can support in this fashion (assuming the object being lifted is in contact with the full area of the mat)?

14-7. A mercury barometer reads standard atmospheric pressure at sea level. When it is carried to an altitude of 5.6 km, the mercury column is half of its original height.
(a) What is the air pressure at this altitude relative to sea-level pressure?
(b) If the barometer is taken up another 5.6 km to an altitude of 11.2 km, will the height of its mercury column fall to zero? Defend your answer.

14-8. A popular classroom demo showing that air occupies space is dunking an empty drinking glass mouth downward in a container of water. Air in the glass prevents water from filling it.

(a) Once submerged, how does the force needed to hold the cup in place change as the cup is pushed deeper into the water?
(b) To what depth must the cup be submerged for the air in the glass to be compressed to half its initial volume?

14-9. A lobster walks onto a bathroom scale on the ocean bottom. The scale, calibrated for normal weight readings in air, shows a weight w.
(a) If the density of the lobster is twice the density of seawater $\left(1025\frac{\text{kg}}{\text{m}^3}\right)$ and the reading on the scale under water is 1.0 N, what is the mass of the lobster?
(b) What is the volume of the lobster?

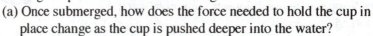

14-10. Pretend that the atmosphere of Earth consisted entirely of water vapor instead of oxygen and nitrogen, with the same atmospheric pressure that Earth now experiences.
    (a) If the water vapor condensed, what would be the depth of the liquid atmosphere? (Hint: Think of a water barometer.)
    (b) If our present atmosphere of nitrogen and oxygen were to condense, would it be less deep or more deep than the depth of a liquid water atmosphere? Nitrogen and oxygen in their liquid phases are respectively 0.8 and 0.9 times denser than water.

14-11. Professor John places a bit of dry ice in an un-inflated balloon. He ties the balloon shut and sets it on a digital balance. As the dry ice sublimes (that is, turns into gas) and the balloon inflates, the reading on the scale goes down.
    (a) What is your explanation for the decreased scale reading?
    (b) If the scale reading decreases by about 2.4 grams, to what volume does the balloon inflate?

14-12. Professor John places a wooden shingle on the lecture table so it overhangs the edge a bit. He covers the shingle with a flattened sheet of newspaper, then strikes the overhanging part of the shingle with a "karate chop" and breaks it.
    (a) The dimensions of the newspaper are 70 cm × 56 cm. What "weight of air" bears down on the paper?
    (b) What role does inertia play in this popular classroom demonstration?

14-13. A balloon is weighted so that it's just barely submerged in water. You poke it a few centimeters deeper beneath the surface of the water.
    (a) Does the balloon sink, remain motionless, or return to the surface?
    (b) What happens to the volume of the balloon at this lower depth?
    (c) Supposing that the balloon sinks, how does depth affect the buoyant force on the balloon?
    (d) How far would it have to sink so the buoyant force on the balloon is half what it was at the surface?

14-14. On a windless morning you hover motionless at low altitude in a hot-air balloon. The total weight of the balloon, including its load and the hot air, is W.
    (a) What is the weight of the displaced air?
    (b) What buoyant force acts on the balloon?
    (c) What is the buoyant force if the total weight of the balloon is 2700 N. (Why do you not need a calculator?)

14-15. A gas-filled balloon rises in air when it weighs less than the air it displaces. It is common to use a gas less dense than air to hold the sides of the balloon out so that it will displace a great amount of air. When filled with helium the balloon's mass is m and its volume is V.
    (a) What is the buoyant force on the motionless balloon?
    (b) How much force is necessary to hold the balloon down?

14-16. The Hindenburg was a German dirigible (also called a Zeppelin) in the late 1930s, filled with bags containing hydrogen. The density of air is 1.29 kg/m^3 and the density of hydrogen is 0.090 kg/m^3. The weight of the empty dirigible was 130 tons. Its volume was 2.01×10^5 m^3.
   (a) What was the buoyant force on the dirigible?
   (b) How much force would be needed to hold it down if it were filled with a hypothetically massless gas?
   (c) What mass of hydrogen would completely fill the interior of the dirigible?
   (d) If completely filled with hydrogen gas, what would have been the net force on the dirigible?

14-17. An empty balloon has a mass m. You wish to fill it with enough helium for it to be able to lift a small package of mass M. The density of helium in the balloon is 0.180 kg/m^3.
   (a) What should be the volume of the balloon to lift the load?
   (b) Calculate the volume if the mass of the balloon is 150 kg and the package to be lifted has a mass of 1000 kg.

14-18. An airplane has a mass m and a total wing area A. During level flight, the pressure on the lower surface of the wings is P.
   (a) What is the pressure on the upper surface of the wings?
   (b) Will the air pressure on the lower surface of the wings increase, decrease, or remain the same when the angle of attack is slightly increased?
   (c) How do the pressure differences on the upper and lower wing surfaces compare when the airplane is descending at a steady rate?

14-19. Water flows at speed v through a fire hose of cross-section A. The fire hose ends in a nozzle of smaller cross-section B.
   (a) What is the speed of the water that exits the nozzle?
   (b) How does the speed of the water leaving the nozzle change if the nozzle is made smaller?
   (c) Calculate the exit speed of the water if the speed of water in the main hose is 4.0 m/s and the cross-section at the nozzle is one-fifth the area of the hose.

14-20. Air flows through a horizontal pipe, flowing first through a section of area A and then through a narrower part of area $0.40A$.
   (a) Compare the relative speeds of air in the wide and narrow sections of the pipe, assuming that air density doesn't change appreciably.
   (b) In which part of the pipe is air pressure greater?

Show-That Problems

14-21. We usually don't notice atmospheric pressure, even though it's quite enormous.
Show that the total weight of air pushing down on a sheet of newspaper of area 0.200 m^2 is 20,200 N.

14-22. A party balloon is squeezed to 2/3 of its initial volume.
Show that the pressure in the balloon increases 1.5 times.

14-23. A suction cup on a ceiling is used to support a 70-kg student.
Show that the contact area between the cup and the ceiling must be a minimum of 0.007 m^2.

14-24. A spherical 5.0-kg balloon filled with helium (density 0.179 kg/m^3) and of volume 500.0 m^3 carries a load and floats motionless in air (density 1.29 kg/m^3).
Show that the mass of the load is 551 kg.

14-25. A liter of liquid air at standard temperature and pressure has a mass of about 0.9 kg.
Show that when it turns to its gaseous phase its volume will expand about 700-fold.

14-26. Average atmospheric pressure at Earth's surface is 1.01×10^5 N/m^2. Earth's radius is 6.37×10^6 m.
Show that the total weight of Earth's atmosphere is about 5.15×10^{19} N.

14-27. In Paul Doherty's physics classes at the Exploratorium, atmospheric pressure is nicely illustrated with a 1.31-m (4.3-foot) bar of steel with a cross-section of 1 square inch. The bar weighs 14.7 pounds, and when it is placed upright on the palm of your hand you can appreciate the magnitude of atmospheric pressure. (Atmospheric pressure is 14.7 lb/in^2 at sea level.)
Show that to simulate atmospheric pressure on Mars, only a wafer 0.8 cm thick does the trick. (Atmospheric pressure on Mars is 0.006 that of Earth's.)

14-28. Atmospheric pressure on the surface of Venus, on the other hand, is much greater: 90 times that of Earth's. Simulating this pressure requires a much longer bar than the 1.31-m bar for Earth.
Show that the length of a bar to simulate the atmospheric pressure of Venus would need to be 118 m tall (which couldn't be held by your hand!).

14-29. Instead of using such a long bar to simulate atmospheric pressure on Venus, the cross-section of the Earth bar at the very end can be reduced to show the same greater pressure (which on the palm of your hand would be dangerous to try holding).
Show that the area at the end of the "Earth bar" should be 1/90th of a square inch.

14-30. Atmospheric pressure on Saturn is 1.4 times that of Earth's.
Show that a Doherty bar to demonstrate Saturn's atmospheric pressure would need to be 1.83 m tall.

14-31. Oil flows at a speed of 1.20 m/s through a pipeline with a radius of 0.30 m.
Show that about 30,000 cubic meters of oil flow in one day.

14-32. Water in a pipe flows into a section that has one-quarter the cross-sectional area of the initial pipe.
Show that its speed increases four-fold.

14-33. The speed of blood in a normal aorta is 0.40 m/s.
Show that if the cross-sectional area of the aorta is enlarged by 1.6, the same volume of blood will flow at a speed of 0.25 m/s.

14-34. A 2% difference in atmospheric pressure exists between the underside and top of a roof during strong winds.
Show that the resulting lifting force on a roof that has a projected horizontal area of 2000 m^2 is more than 4 million N.

14-35. Relative to an airplane, the speed of air over the plane's wing is 63 m/s and under its wing is 60 m/s. Suppose that the plane has a wing area of 24 m^2 and that the density of air at the airplane's flying altitude is 1.13 kg/m^3.
Show that the lifting force is about 5000 N.

15 Temperature, Heat and Expansion

nply put, temperature is a measure of the average translational KE of the molecules in a ostance. Raising the temperature of a substance indicates increasing the average speed of the olecules within it.

e amount of energy involved in a temperature change depends upon three factors:

1. The nature of the substance involved. All else being equal, some materials require more energy per gram than others to raise their temperature by 1°C. This "amount of energy to raise the temperature of 1 gram of something by 1°C" is called the *specific heat capacity* of the material, and is given the symbol c. It has units of J/g·C° or kJ/kg·C°.*

2. The mass of the sample, m. (It takes more energy to raise the temperature of a substance with more molecules.)

3. The number of degrees of temperature change, ΔT. (A larger change in temperature requires more energy than a smaller temperature change).

l of the above apply equally well to objects cooling or objects warming.

e can summarize the above factors as follows:

Heat involved in a temperature change	=	Specific heat capacity of the substance	×	mass of stuff changing temperature	×	Number of degrees change in temperature

$Q = c\,m\,\Delta T$. (Q is the conventional symbol for "quantity" of heat.)

materials warm, their particles jiggle faster and move farther apart, on average. The result is expansion of the material. The amount of expansion depends on the type of material and the nperature change. For the same 1 degree temperature change, some materials will expand more in others. The fractional change in length ($\Delta L/L$) per degree is called the *coefficient of linear oansion* and is given the symbol α, with units 1/C°.

So $\dfrac{\Delta L}{L} = \alpha \Delta T$ or $\Delta L = \alpha L \Delta T$.

e say that something's change in length depends on the material of which it is composed hich determines α), its initial length, and the number of degrees it is warmed or cooled.

---

listorically, people took equal masses of hot stuff (for example 10 grams of water, iron, lead, etc. at 0°C) and placed them on a chunk of ice. Each hot sample melted a different amount of ice. Whichever nple melted the most ice had the largest "heat capacity." Water was assigned a specific heat capacity of 'calorie"/g·C° and all of the other materials were assigned heat capacities based on how much ice they ild melt compared with an equal amount of water. Today it is customary to use joules rather than ories.

It doesn't make sense to speak of a change in length for liquids. We instead speak of a fractional change in volume per degree, called the *coefficient of volume expansion*, which is given the symbol β, also with units $1/C°$. Analogous to linear expansion we have the equation

$$\Delta V = \beta V_0 \Delta T.$$

Some useful things to know in solving these problems:
1 milliliter of water has a mass of 1 gram.
1 liter of water has a mass of 1 kg.
1 calorie = the amount of energy needed to raise the temperature of 1 gram of water by $1\ C° = 4.18\ J$.
1 Calorie = 1 kcal = 1000 calories.

Material	c, Specific heat capacity $(J/g \cdot C°$ or $kJ/kg \cdot C°)$
Water	4.18
Aluminum	0.900
Clay	1.4
Copper	0.386
Lead	0.128
Olive Oil	1.97
Silver	0.23
Steel (iron)	0.448

Material	α, Coefficient of Linear Expansion $(1/C°)$
Steel	11×10^{-6}
Brass	19×10^{-6}
Aluminum	23×10^{-6}
Glass	10×10^{-6}
Gold	14.3×10^{-6}

Sample Problem 1

Joan wishes to take a hot bath. She fills the tub with x liters of water at a temperature T_{hot}.
(a) If the initial temperature of the water was T_{cold}, how much energy went into heating the water? (1 liter of water has a mass of 1 kg).

Focus: $Q = ?$
The heat involved changing temperature depends on the specific heat capacity of the water, c_w the mass of water, and the temperature change $T_{hot} - T_{cold}$. Since 1 liter of water has a mass of 1 kg, m (in kg) $= x$ (in liters) $\times \dfrac{1\,\text{kg}}{1\,\text{liter of water}}$. So

$$Q = c\, m\, \Delta T = c_w x (T_{hot} - T_{cold}).$$

(b) Calculate the amount of heat required to raise the temperature of 320 L of water from 18°C to 52°C.

$$Q = c_w x (T_{hot} - T_{cold}) = \left(4.18 \tfrac{kJ}{kg \cdot C°}\right)\left(320\,L \times \frac{1\,\text{kg}}{1\,L\ \text{of water}}\right)(52°C - 18°C)$$

$$= 45,000\ \text{kJ}.$$

ample problem 2

rmon has taken the 2nd place medal at the County ir for his collection of belly-button lint, but he ibts that his medal is really silver. He takes the dal, mass m_m, and dunks it into a pot of boiling ter in order to heat it to 100 °C. He puts a mass of ter m_w into a Styrofoam cup and measures its ial temperature, T_0. He removes the medal from boiling water and puts it into the Styrofoam cup. er stirring, the water in the cup ends up at T_f.

What is the specific heat capacity of his medal?

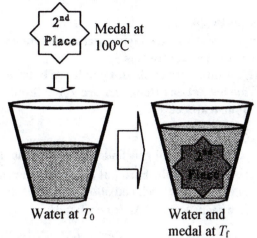

Water at T_0 | Water and medal at T_f

Focus: $c_m = ?$

Harmon's method depends on the law of conservation of energy as applied to heat transfer. Whatever energy the medal gives up, Q_{medal}, is gained by the water, Q_{water}.[*] That is $-Q_{medal} = Q_{water}$. We say Q_{medal} is negative because heat flows *from* it, while Q_{water} is positive because heat flows *into* it.

From $-Q_{medal} = Q_{water}$ $\Rightarrow -c_m m_m \Delta T_m = c_w m_w \Delta T_w$ $\Rightarrow c_m = \dfrac{-c_w m_w \Delta T_w}{m_m \Delta T_m} = \dfrac{-c_w m_w (T_f - T_0)}{m_m (T_f - 100°C)}$

Harmon merely has to compare the value for his calculated specific heat capacity with the specific heat capacity of silver.

Harmon's medal has a mass of 68 grams, the cup initially holds 155 grams of water at 21.0°C, and the final temperature of the water and medal is 24.5°C. Determine whether or not his medal is silver.

Solution: $c_m = \dfrac{-c_w m_w (T_f - T_0)}{m_m (T_f - 100°C)} = \dfrac{-(155\,g)(24.5°C - 21.0°C)\left(4.18\frac{J}{g\cdot C°}\right)}{(68\,g)(24.5°C - 100°C)} = 0.44\frac{J}{g\cdot C°}.$

This result is far from the specific heat capacity of silver (0.23 J/g·C°), so the medal is definitely not silver. What metal might it be made of?

---

this problem and in all the others in this chapter we assume that the system is *closed*—that is, no rgy leaves or enters the system (we approximate this condition by doing the experiments in a Styrofoam). We also assume that the system has reached *equilibrium*—that is, that the water and the medal both e exactly the same temperature throughout. We're also assuming that no hot water "takes a ride" on the lal when we transfer it to the cup. Careful technique in the lab can help one *approach* these ideal imed conditions.

Sample Problem 3

Suppose that you are laying sections of steel railroad track L meters long on a cool day where the temperature is T_c.

(a) How much space should you leave between sections of track so that they can expand on the hottest day (temperature T_h) without buckling? (See Figure 15.14 on page 298 in your textbook.)

Gap = ?

As each steel rail gets hotter it expands into the gap between the rails. Each rail expands by an amount $\Delta L/2$ into the gap on each end of the rail. Thus the total gap size needs to equal ΔL for each rail.

Gap = $\Delta L = L_{rail}\alpha_{steel}\Delta T_{rail} = L_{rail}\alpha_{steel}(T_h - T_c)$.

(b) Calculate the size of the gap for 18-meter long steel rails laid down on a day when it's 9°C outside, where the hottest day on record is 42°C.

Gap = $\Delta L = L_{rail}\alpha_{steel}(T_h - T_c) = (18\text{m})\left(11\times10^{-6}\tfrac{1}{C^\circ}\right)(42^\circ\text{C} - 9^\circ\text{C}) = 0.0065\text{m} = \textbf{6.5 mm}$.

Problems

15-1. People in the pioneering days placed hot potatoes in their pockets on cold winter days to keep their hands warm.
   (a) Assuming that a potato is mostly water, how much energy could be released by a hot potato of mass m and temperature T_0 in cooling to T_f?
   (b) Calculate the heat released by a 350-g potato that cools from 85°C to 15°C.

15-2. Your thin plastic water bottle holds m grams of water at temperature T_0. You put it into the refrigerator to cool it to temperature T_f.
   (a) How much heat does the refrigerator remove from the water?
   (b) Calculate how much heat the refrigerator removes from 500 mL of water in cooling it from 28°C to 4°C.

15-3. An amount of energy Q raises the temperature of some water from T_0 to T_f.
   (a) What mass of water is being heated?
   (b) Calculate the mass of water if 21 kJ changes it from 0°C to 100°C.

15-4. You decide to try the "Ice-Water Diet" and drink water at 0°C to warm up to your body temperature, 37°C.
   (a) How much ice water must you drink to "burn" 3500 Calories (the approximate energy content of one pound of fat)? (Each Calorie is 1000 calories.)
   (b) Why would you recommend, or not recommend this diet to overweight friends?

15-5. An amount of heat Q raises the temperature of m grams of metal from T_0 to T_f.
   (a) Calculate the specific heat capacity of the metal.
   (b) Metal having a mass of 48.4 grams experiences an increase in temperature from 34.0°C to 47.5°C when 252 J of heat is added to it. Calculate the specific heat capacity of the metal.

15-6. Equal amounts of energy are added to equal masses of two different metals. The first metal experiences a temperature change of ΔT_1.
 (a) How does the temperature increase of the second metal compare with the temperature increase of the first metal?
 (b) A certain amount of heat added to 65 grams of lead causes a temperature change of 15.0°C. If the same amount of energy were added to 65 grams of copper, by how much would its temperature change?

15-7. You want to find the specific heat capacity of some vegetable oil. You place a mass of oil m_{oil} into a Styrofoam cup at a temperature T_0 and add a mass m_{Al} of hot aluminum at a temperature T_h to the oil. The final temperature of the two is T_f.
 (a) What is the specific heat capacity of the oil?
 (b) Aluminum with a mass of 57 grams at 100°C is added to 65 g of oil at 23°C in a Styrofoam cup. The combination is stirred, and the final temperature is 46°C. Calculate the specific heat capacity of the oil.

15-8. A copper block of mass M at temperature T_{cop} is placed into an insulated cup containing a mass m_w of water at a temperature T_0.
 (a) To bring the final temperature of the water to T_f, what should be the initial temperature of the copper?
 (b) Water with a mass of 250 g at 24°C is in a Styrofoam cup. Calculate the initial temperature of a 350-g copper block that will produce a final temperature of 21°C.

15-9. Your (perfectly insulated) bathtub has x liters of water in it, but it has cooled down to a temperature T_c. You'd like to add just the right amount of hot water at temperature T_h to make your bathwater the perfect temperature T_p.
 (a) How much hot water must you add? (Recall that 1 liter of water has a mass of 1 kg.)
 (b) You like your bath to be at 42°C, but the 82 liters of water presently in the tub are at 35°C. Calculate the amount of hot water at 50°C you must add to the tub.

5-10. You pour x mL of hot coffee at T_h into a steel coffee cup of mass m at an initial temperature of T_c.
 (a) To what temperature does the coffee immediately cool?
 (b) Calculate the final temperature of 250 mL of coffee at 87°C after it is poured into an 85-g steel coffee cup originally at 22°C.

5-11. Your half-finished cup of coffee has cooled down to a temperature T_c. You like your coffee to be at the perfect temperature T_p. You put your cup, containing x mL of coffee, into the microwave oven. (One mL of coffee has a mass of 1 gram.)
 (a) How much energy will it take to restore your coffee to its "perfect" temperature? Assume that the coffee has the same thermal properties as water and that the cup itself gains negligible heat from the microwave oven.
 (b) The oven delivers energy to the coffee at a rate of P watts (P J/s). How much time should you set the microwave to heat your coffee to the perfect temperature?
 (c) If the microwave oven delivers 750 W to the coffee, how long will it take to reheat 140 mL of coffee from 22°C to 83°C?

15-12. A bomb calorimeter consists of a sealed stainless steel container (the "bomb") into which is placed a sample to be burned. The container is filled with oxygen under pressure, and the whole thing is surrounded by water in an insulated container. When a small fuse ignites the sample the temperature of the water (and the bomb) increases, from which you can determine how much energy was released.

You decide you want to know how many Calories there are in peanuts. You place x grams of peanuts into a steel bomb of mass m_b and surround the bomb by a mass m_w of water. Then you ignite the fuse.

(a) If the temperature of the water and the steel bomb increases by ΔT, how much energy did the peanuts release? (Assume that all of the released energy goes into the bomb and the water. The small part of the energy that goes into heating the products of the combustion can be ignored.)

(b) If the energy to warm the water and the steel bomb came from x grams of peanuts, how much energy was provided by each gram of peanuts?

(c) Suppose that you use 0.95 g of mashed peanuts, the mass of the bomb is 750 g, the mass of the water is 450 g, the initial temperature of the water is 19.1°C and the final temperature is 29.4°C. What is the calorific value of peanuts in Cal/g? (Remember that 1 Calorie = 1,000 calories = 4,180 J.)

15-13. A light bulb of wattage P is placed into an insulated cup filled with a mass m of oil with a specific heat capacity c_{oil}. You turn on the light bulb. The oil begins to get hotter.

(a) How much heat does the bulb transfer to the oil in a time t?

(b) How much warmer will the oil be after the bulb has been submerged in it for a time t?

(c) Calculate the temperature change of 150 g of olive oil when a 60-watt bulb is submerged in it for 45 seconds.

15-14. Some homes have "on-demand" water heaters. Rather than storing hot water in a large tank these heaters activate when you turn the hot water on, and provide hot water only as long as it is needed. In taking a shower, suppose that you use x liters of hot water each minute (recall that each liter of water has a mass of 1 kg.)

(a) How much energy is required to raise the temperature of x liters of water from T_{cold} to T_{hot}?

(b) How much energy is required per second (that is, how much power) to heat x liters of water from T_{cold} to T_{hot} in a time interval of 1 minute?

(c) Calculate the power rating for a perfectly efficient electric heater designed to heat 3.0 liters of water from 15°C to 50°C each minute.

15-15. A nuclear power plant typically produces 2 J of heat for every joule of electric energy that it produces. The heat is often removed by water pumped through a secondary cooling loop. Suppose that your power plant produces x MW (megawatts, or million watts) of electric power and is next to the ocean. The water your reactor takes in and discharges back into the ocean is limited to a temperature rise of no more than ΔT_{max}.

(a) How much heat does the reactor release every second?

(b) How much heat can m_w kilograms of water absorb if it is going to increase its temperature by ΔT_{max}?

(c) How much cooling water must you take into your reactor each second?

(d) How much cooling water per second is required to cool a 1000 MW reactor if we stipulate that the cooling water can only have a maximum temperature rise of 4°C?

15-16. • A solar water heater has a collector plate of area A (in units m^2). It receives solar energy at a rate I_0 (watts/m^2) and transfers a fraction ε of that energy to water passing through the collector plate at a rate R (mL/second). (The fraction ε is called the efficiency of the collector.)
   (a) How much energy does the collector receive in one second?
   (b) If the collector delivers the collected solar energy to the water with an efficiency of ε, how much energy does the water gain each second?
   (c) If R mL/s (R grams/s) of water pass through the collector, by how many degrees does the water heat up?
   (d) Calculate the change in temperature for water that runs through a 1.2 m^2 solar collector at a rate of 15 mL per second at a time of day when the sunlight hits the collector with an intensity of 620 W/m^2, and transfers that energy to the water with an efficiency of 48% ($\varepsilon = 0.48$).

15-17. Two identical blobs of clay of mass m moving at speed v have a head-on collision. Both stick together.
   (a) Use conservation of momentum to calculate the final speed of the combined blobs of clay.
   (b) What was the initial KE of the two-clay-blobs system?
   (c) What is the final KE of the two-clay-blobs system?
   (d) By how much did the KE of the two-clay-blobs system change?
   (e) Assuming that all of the "lost" KE is converted to heat within the clay itself, by how much does the temperature of the clay blobs increase?
   (f) Calculate the increase in temperature for two colliding 12-g clay blobs each initially moving at 5 m/s.

15-18. A car of mass M moving at a speed v applies its brakes and comes to a stop. The car has disk brakes on all four wheels, four disks total. Each of the brake disks has mass m.*
   (a) By how much does the temperature of the brake disks increase due to slowing of the car? (Assume that all of the KE of the car goes into heating the brake disks.)
   (b) Calculate the increase in temperature of the disk brakes of a 1200 kg car that brakes from 25 m/s to rest in 15 seconds if each of the four brake disks is made of steel and has a mass of 5.5 kg.

15-19. A mass m of lead shot at a temperature T_0 is put into a cardboard tube of length L. Both ends of the tube are stoppered. The tube is inverted n times. With each inversion the lead falls through the distance L. Assume that all the PE of the lead goes into warming the lead shot.
   (a) What is the final temperature of the lead?
   (b) Calculate the final temperature of 25 grams of lead in a tube 1.0 m long after the tube has been inverted 50 time. The initial temperature of the lead is 22.0°C. (Hint: 1 J = 1 kg·m^2/s^2. Be sure to express quantities in consistent units.)

lead
shot

---

* disk brakes, a metal disk rides inside the rim of each wheel. Brake pads on either side of the disk ⌐eeze on the disk to slow the car. Friction between the pads and the disk warms the brake disks.

15-20. A rod of length L is heated from a temperature T_0 to a temperature T_f.
   (a) By how much does the length of the rod increase?
   (b) Calculate the increase in length of a 0.901 m long steel rod that is heated from 20°C to 100°C.

15-21. A beam of length L becomes shorter by a distance ΔL when it goes from temperature T_0 to a lower temperature T_f.
   (a) What is the coefficient of linear expansion of the rod?
   (b) Calculate the coefficient of linear expansion for an 8.20-m long beam that contracts by 2.4 mm when it cools from 27°C to 16°C.

15-22. Before going to bed you remove your 24-karat gold ring (of inner diameter d) from your warm finger (at T_h) and set the ring on the nightstand. The following morning when the bedroom temperature is T_{br} you attempt putting on the ring.
   (a) By how much has the diameter of the ring shrunk overnight?
   (b) Calculate the change in diameter of the ring for an initial diameter of 1.90 cm, a finger temperature of 36°C, and a bedroom temperature of 20°C.

15-23. A tower h meters tall is made of iron.
   (a) By how much does its height increase as the temperature increases from T_0 to T_f?
   (b) The Eiffel Tower in Paris is made of cast iron and is approximately 300 m tall. By how much does its height increase between a 5°C morning and a 19°C afternoon?

15-24. Water in a cylindrical reservoir of radius R and average depth D has an average temperature T.
   (a) If the temperature of the reservoir were to increase by 10°C, by how much would the water level rise due to thermal expansion? (For simplicity, assume the reservoir itself doesn't expand.)
   (b) Calculate the approximate rise in level for a cylindrical reservoir of diameter 40 m and a depth of 20 m when water temperature increases by 10°C.
   Assume that $\beta_{water} \approx 210 \times 10^{-6}/\text{C}°$.

15-25. A popular demonstration of linear expansion involves trying to place a brass ball of diameter $d_{0(ball)}$ through a brass ring of a slightly smaller diameter, $d_{0(ring)}$. The initial temperature of the ball and ring is T_0. The ring is placed in a flame and heated until it has expanded enough for the ball to pass through.
   (a) What is the minimum temperature T_f to which the ring must be heated before the ball will fit through it?
   (b) Calculate the value for T_f for a 2.540 cm diameter ball and a ring with an inner diameter of 2.530 cm, both with an initial temperature of 21°C.

5-26. • Suppose that NASA tool designers want to design a
steel wrench that will be used by astronauts to tighten or
loosen a d-mm brass nut on the outside of their lunar
lander. The challenge is to design and manufacture a
wrench here on Earth at 20°C where it will be a little bit
too big, but will fit just right at a higher temperature when
both the nut and the wrench have expanded. The temperatures on the lunar surface can
reach a maximum temperature T_{max}.

(a) Write an equation for the width of the brass nut at $T = T_{max}$.
(b) Let $d_{0(wrench)}$ be the width of the wrench at 20°C. Write an equation for the width of
the wrench at $T = T_{max}$.
(c) To have the wrench and the nut the same size at T_{max}, how big does the gap in the
wrench have to be at 20°C?
(d) The maximum temperature on the Moon's surface is 100°C. The width of the brass
nut at 20 °C is 12.00 mm. Calculate the width of the "gap" in the wrench at 20°C if it
is going to fit onto the brass nut when both of them are at 100°C.

Show-That Heat and Temperature Problems

15-27. A 62.3-gram sample of aluminum is heated.
Show that 516 J will raise its temperature by 9.2°C.

15-28. A brass rod is 2.4 m long at 21°C.
Show that it will be 3.6 mm longer at 100°C.

15-29. A heat input of 35.9 J raises the temperature of 47.0 grams of a material by 2.5 C°.
Show that the specific heat capacity of the material is 0.306 J/g·C°.

15-30. A piece of silver of mass 88.9 g at 93°C is placed in 175 g of olive oil at 18.0°C in an insulated container.
Show that the final temperature will be 22.2°C.

15-31. A certain amount of heat raises the temperature of a sample of iron by 10°C.
Show that the same amount of heat will raise the temperature of an equal mass of lead by 35°C.

15-32. Suppose that you want to place an aluminum rod 2.000 cm in diameter into a round hole 1.998 cm across.
Show that you have to lower the temperature of the aluminum rod by 44°C to get it to fit.

15-33. A 30.0-m length of steel rod expands 8.00 mm when heated.
Show that the temperature change is 24 C°.

15-34. Suppose that at 10°C you have a steel rod that is 1.0065 m long and an aluminum rod that is 1.0049 m long.
Show that both rods will have the same length at 143°C.

15-35. An aluminum rod grows 0.0033 m in length when its temperature is raised from 10°C to 90°C.
Show that its initial length before being heated was 1.8 m.

16 Heat Transfer

Conduction

Consider a copper rod wrapped in insulation and running between a container of boiling water and a container of ice water.

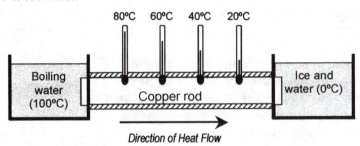

80°C   60°C   40°C   20°C

Boiling water (100°C)

Copper rod

Ice and water (0°C)

Direction of Heat Flow

Heat will flow through the copper rod from the hot side to the cold side. At equilibrium, the rate of heat flow through the rod will depend on several factors:

1) What the rod is made out of—some materials conduct heat better than others. How well a material conducts heat is given by its *coefficient of thermal conductivity*, κ, which has units of W/m·C° (the units will make sense below).
2) The cross-sectional area of the rod, A (in unit m^2)—doubling the cross sectional area is just like having two rods to conduct heat instead of just one.
3) The temperature difference across the ends of the rod, ΔT (in C°)—a larger temperature difference across the rod will cause more heat to flow through the rod.
4) The length of the rod, L (in unit m)—if the rod were twice as long, each half of the longer rod would have half the temperature difference across the initial rod. The rate of heat flow would be half:

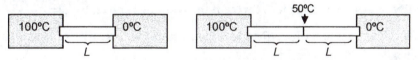

100°C   0°C   100°C   50°C   0°C

L   L   L

In this example, the temperature difference over a length L is 100 C° for the first rod, but only 50°C for the second rod of length L, so the rate of heat flow is half as much. The rate of heat flow is inversely proportional to the length of the rod.

In summary, the heat Q conducted during a time t through a bar of length L and cross-sectional area A is $Q = \dfrac{(\kappa A \Delta T)}{L} t$, where ΔT is the temperature difference between the ends of the bar and κ is the thermal conductivity of the material. The SI unit of thermal conductivity is J/(s·m·C°), or equivalently, W/m·C°.

145

Sample Problem 1

A glass window in a house has a height h, width w, and a thickness L. The temperature inside the house is T_{in}, while the lower temperature outside is T_{out}.

(a) What quantity of energy transfers through the window by conduction in time t?

Focus: $Q = ?$ Heat will flow through the window from the warmer to the cooler side.

$$Q = \frac{(\kappa A \Delta T)t}{L} = \frac{\kappa_{glass}\, wh\,(T_{inside} - T_{outside})t}{L}.$$

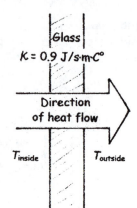

Glass
$\kappa = 0.9 \text{ J/s·m·C}°$

Direction
of heat flow

T_{inside} $T_{outside}$

(b) A window in a house is 63 cm wide and 84 cm high with glass 4.0 mm thick. The temperature inside the house is 22°C, while outside it is 4°C. Calculate the quantity of heat that flows due to conduction through the window in one hour. Thermal conductivity κ for the glass is 0.9 J/s·m·C°.

$$Q = \frac{\kappa_{glass}\, hw\,(T_{in} - T_{out})t}{L}$$

$$= \frac{\left(0.9\frac{J}{s\cdot m\cdot C°}\right)\left[\left(84\,cm \times \frac{1m}{100\,cm}\right)\times\left(63\,cm \times \frac{1m}{100\,cm}\right)\right](22°C - 4°C)(3600\ s)}{\left(4.0\,mm \times \frac{1m}{1000\,mm}\right)} = 7.7 \times 10^6\ J.$$

(c) How would the amount of heat loss differ if the glass were 5.0 mm thick instead of 4.0 mm thick?

Answer: For a larger L the quantity of heat is smaller. Doing the same calculation as in (b), with $L = 5.0$ mm, gives 4/5 the previous answer—**0.80 times as much.**[*]

(d) The quantity of heat that transfers through the window is given by $Q = \dfrac{(\kappa A \Delta T)t}{L}$.

How would this expression differ for the *rate* at which this quantity of heat energy flows through the glass?

Answer: The rate of energy flow per unit time would be $\Delta Q/\Delta t$, or simply the above expression divided by time t. Then we have

$$\frac{Q}{t} = \frac{(\kappa A \Delta T)}{L}.$$

---

[*] Actual heat transfer is more complicated than this. As heat flows through the window a thin layer of air on the cold side of the window warms and a thin layer of air on the warmer side of the window cools. Since total ΔT will be across the window *plus* a layer of air on either side of the window (and air is a poor conductor of heat) the rate of heat flow will drop. Meanwhile, convection will also kick in as the warmer air will tend to rise and be replaced by colder air. In addition, the warmer parts of the glass will be losing energy by radiation to the cooler outdoor environment, and the room is radiating some of its heat out the window! What we are working on here is a first, *simple* model of heat transfer. The reality is more complicated, but comprehensible. If you continue in your study of heat there'll be plenty more to learn!

(e) Calculate the rate at which heat flows through the glass window.

Solution:

$$\frac{Q}{t} = \frac{7.7 \times 10^6 \, J}{3600 \, s} = 2100 \, W.$$

Radiation

Radiant energy is in the form of electromagnetic waves (light, for example). The peak frequency at which a body radiates is proportional to the body's surface temperature. In the textbook we express this as $\bar{f} \sim T$, where $\bar{f}$ is the peak frequency of radiation, and T is the temperature of the emitter in kelvins. We go further here and express this relationship as an equation, called *Wien's Law*:

$$\bar{f} = \left(5.88 \times 10^{10} \, \tfrac{Hz}{K}\right) T$$

Again, T is measured in kelvins. Even at relatively modest temperatures the frequency is going to be a huge number. The frequency of visible light ranges from about 4.3×10^{14} Hz (waves per second) for red light to 7.5×10^{14} Hz for violet light.

Sample Problem 2
The red supergiant Antares has a peak radiation frequency of 2.0×10^{14} Hz.

(a) What is the approximate surface temperature of Antares?

From $\bar{f} = \left(5.88 \times 10^{10} \, \tfrac{Hz}{K}\right) T \Rightarrow T = \dfrac{\bar{f}}{5.88 \times 10^{10} \, \frac{Hz}{K}} = \dfrac{2.0 \times 10^{14} \, Hz}{5.88 \times 10^{10} \, \frac{Hz}{K}} = \textbf{3400 K.}$

(b) What would be the peak frequency of radiation from Antares if it had twice the temperature?

Answer: From our relation $\bar{f} \sim T$, twice the temperature would be twice the frequency. Twice 2.0×10^{14} Hz = **4.0×10^{14} Hz.**

$$\bar{f} = \left(5.88 \times 10^{10} \, \tfrac{Hz}{K}\right) T = \left(5.88 \times 10^{10} \, \tfrac{Hz}{K}\right) \times 2(3400 \, K) = \textbf{4.0} \times \textbf{10}^{14} \, \textbf{Hz.}$$

Problems

16-1. Q joules of heat flow each second through a copper rod of cross-sectional area A and length L.
    (a) What temperature difference must exist between the two ends of the rod?
    (b) Calculate the temperature difference between the two ends of the rod. You need to know that $\kappa_{copper} = 401$ W/m·C°, 125 joules flow each second, the rod's length is 12 cm and its diameter is 1.26 cm.

16-2. Suppose that the air temperature right below the plaster ceiling of your house is a warm T_0, but your (uninsulated) attic is at a chillier T_a.
    (a) If the area of your home's ceiling is A and the plaster has a thickness L, at what rate is heat being lost through ceiling?
    (b) Suppose that the air temperature just below the ceiling is 23°C and the temperature of the air in the attic is 5°C. Calculate the rate of heat loss for a 1.6-cm thick plaster ceiling having an area of 110 m². Assume $\kappa_{plaster} = 0.21$ W/m·C°.

16-3. A pond in winter has a layer of ice of thickness x. The temperature of the air above the pond is T_a, while the temperature of the water directly below the ice is 0°C.
    (a) How much energy passes through each square meter of ice each second?
    (b) If the temperature outside is -10°C, how much heat passes through each square meter of ice each second? Assume the ice is 5.0 cm thick and that $\kappa_{ice} = 2.2$ W/m·C°.
    (c) A 1-m² layer of ice 1 mm thick has a mass of 0.916 kg. Removal of 335,000 joules of energy from the water at 0°C forms 1 kilogram of ice. If the temperature remains -10°C outside, how long will it take for the ice to get 1 mm thicker?
    (d) What happens to the rate of ice formation as the ice grows thicker?

16-4. A typical lab experiment for measuring the coefficient of thermal conductivity of a metal bar involves taking an insulated bar of length L and diameter d and placing it between two insulated chambers, similar in setup to the first diagram in this chapter, page145. In one chamber steam at 100°C is blown onto the hot side of the bar. In the other chamber cold water with an initial temperature T_0 flows past the other side of the bar and leaves at a slightly higher temperature T_f. The average temperature the cold side of the bar "sees" is $(T_0 + T_f)/2$. A beaker is used to catch a mass m of cold water flowing past the cold end of the bar in a time t.
    (a) What is the coefficient of thermal conductivity of the bar?
    (b) A brass bar has a diameter of 2.54 cm and a length of 25.0 cm. One end is in 100° steam and the other end encounters incoming water temperature of 22.3°C, with the outgoing temperature being 24.7°C. 102 grams of water flow by the cold side of the bar in 55 seconds. Calculate κ for the bar.

16-5. A block of ice keeps the inside of an ice chest at 0°C. The ice chest is made from Styrofoam of thickness L, a total surface area A, and hangs by thin ropes from a tree.
    (a) If the temperature outside is T, at what rate is the ice chest absorbing energy?
    (b) Suppose that you start out with a block of ice of mass m in the ice chest. If it takes x joules of energy to melt a kilogram of ice, how long will it take for all of the ice in the ice-chest to melt?
    (c) Calculate answers to (a) and (b) for an ice chest 55 cm long, 42 cm wide and 42 cm deep, made with walls 2.5 cm thick, which holds 7.5 kg of ice on a 25°C day. Use $\kappa_{Styrofoam} = 0.026$W/m·C°. 335 kJ are required to melt 1 kg of ice.

16-6. An old laboratory oven is shaped like a cube of length s on each side. The walls and oven door are made from asbestos of thickness a with thin sheets of metal on either side (for all practical purposes, the contribution of the metal sheets to resisting heat flow through the walls can be ignored). The oven has an electric heating element whose maximum power output is P ($\kappa_{asbestos} = 0.20$ W/m·C°).
  (a) Ignoring losses to radiation and convection, what is the maximum temperature that the oven can maintain in a room at T_{room}?
  (b) Suppose that the laboratory oven is 30 cm on each side, the asbestos walls are 1.2 cm thick, and the heating element puts out a maximum 2000 watts. The stove sits in a 20°C room. Calculate the maximum temperature possible inside the stove.

16-7. In a lab experiment a heated plywood box l meters long, w meters wide and h meters deep provides warmth to some biological samples at a constant temperature T_h. The walls of the box have a thickness x.
  (a) If the box is to be kept in a room whose temperature is T_r, at what rate must power be supplied to the box's interior? Ignore radiative and convective losses from the box.
  (b) Suppose that the box is 2.0 m long, 0.60 m wide, and 0.50 deep, and the plywood is 12.7 mm thick. Room temperature is 22°C. How much power must be provided to the interior of the box to maintain its contents at 55°C? Use $\kappa_{plywood} = 0.11$ W/m·C°.

16-8. Referring to the previous problem, suppose that the plywood walls could be replaced by the same thickness of Styrofoam ($\kappa = 0.026$ W/m·C°).
  (a) How much power would have to be supplied to the box to maintain its interior temperature at 55°C?
  (b) What would be the rate of heat flow if the Styrofoam were twice as thick?

16-9. Earth has an average surface temperature of about 18°C.
  (a) What is the peak frequency at which Earth radiates?
  (b) In which part of the electromagnetic spectrum is this terrestrial radiation?

16-10. The tungsten lamp filament in a 60-watt light bulb typically operates at a temperature of 2800 K.
  (a) What is the peak frequency at which the filament radiates?
  (b) In comparison to red light, where in the electromagnetic spectrum does this frequency occur?

Show-That Problems

16-11. Each second 29.1 joules flow through a 22-cm long nickel rod (κ_{nickel}= 91 W/m·C°) when the temperature difference between the two ends of the rod is 56°C.
Show that the diameter of the rod is 4.0 cm.

16-12. A 10 cm thick wall made from concrete (κ = 0.8 W/m·C°) separates two sections of a flowing stream. On one side the water is 18°C and on the other side it is 23°C.
Show that heat flow through each square meter of wall is 40 J/s.

16-13. A 2.5 cm diameter rod made from aluminum ($\kappa \approx 240$ W/m·C°) runs between a 56°C temperature bath and a 42°C temperature bath. Heat flow through the rod is 33 W.
Show that the length of the rod is 5.0 cm.

16-14. The dim red dwarf star *Wolf 359* has a surface temperature of 2430 K and is 7.8 light-years from Earth.
Show that the peak frequency at which *Wolf 359* radiates is 1.43×10^{14} Hz.

16-15. A star has a surface temperature of approximately 5800 K.
Show that the peak frequency of the light radiated by the star is 3.4×10^{14} Hz.

17   *Change of Phase*

As the particles in a substance gain energy they vibrate faster. The temperature of the substance rises in proportion to the quantity of heat added, the mass of the substance, and its specific heat capacity. Recall from Chapter 15 that the *specific heat capacity* of the substance is given the symbol c and has units J/g·C° or kJ/kg·C°. Note that the numerical value of c is the same in either of these alternate units.

At some point in the heating process the particles may move fast enough to overcome the forces bonding them together. A solid can then change phase to a liquid and a liquid to a gas. For a solid, any energy input at the phase-change point goes into separating the molecules from one another. This process is called *melting*. During melting, and also during vaporization, heat is being added to the substance without increasing its temperature. The added heat is called the *latent heat* and is given the symbol L.

The quantity of heat involved in a phase change from solid to liquid (or from liquid to solid) is called the *latent heat of fusion*, L_{fus}. For the similar process of going from liquid to vapor (or vice versa) the heat involved is called the *latent heat of vaporization*, L_{vap}. (The heat involved in the process of going from solid directly to vapor is called the *latent heat of sublimation*, L_{sub}.)

Consider a glass of water placed in a temperature-controlled room at exactly 0°C. When the water cools to 0°C it won't freeze. That's because in order to freeze, heat must be removed from the water. For this to occur, the environment must be *colder* than 0°C.

Likewise, consider placing a very cold ice cube into the same temperature-controlled room. The ice cube will warm to 0°C. It won't melt because melting requires heat input, which means the surroundings would have to be at a higher temperature than 0°C. So when we say that ice melts at 0°C, or water freezes at 0°C, we mean that 0°C is the temperature of the ice *when* it melts or of the water *when* it freezes, rather than meaning that a change of phase automatically occurs when the temperature of the environment is or becomes 0°C.

Conservation of energy rules when doing heat problems. We note the initial state of a system, the final state of that system, and calculate the quantity of energy involved in going from the initial state to the final state. If the system is *closed* (no energy enters or leaves the system) then the total energy in the system simply changes form. In this view, temperature is akin to kinetic energy (and actually *is* kinetic energy per molecule in the system), and latent heat is akin to potential energy.

Suppose that we place a very cold ice cube into a warm room and want to know how much energy is involved in the ice cube melting and the resulting water warming up. We know, physically, that the ice on the surface of the cube will begin melting before the center of the ice cube has warmed to 0°C. But we can think of this melting and warming process as occurring in discrete steps.

1. Very cold ice warms up to ice at 0°C $\Rightarrow Q$ (our symbol for quantity of *heat*) $= c_{ice} m_{ice} \Delta T_{ice}$
2. Ice at 0°C melts to water at 0°C $\Rightarrow Q = m_{ice} L_{fus}$
3. Water at 0°C warms up to water at $T_f \Rightarrow$
   $$Q = c_{water} m_{ice} \Delta T_{ice\ water} = c_{water} m_{ice} (T_f - 0°C) = c_{water} m_{ice} T_f \ (T_f \text{ in Celsius degrees}).$$

The total heat involved in the process is the sum of these three quantities.

Whereas masses have been expressed in kilograms in previous chapters, we now find it more convenient to make use of the gram (where $1\ g = 10^{-3}\ kg$), since we deal mainly with smaller amounts of ice and water.

Some useful numbers for solving problems in this chapter:
L_{fus} (water) $= 335\ J/g$
L_{vap} (water) $= 2260\ J/g$
$c_{ice} = 2.09\ J/g\cdot°C$
$c_{water} = 4.18\ J/g\cdot°C$

Other specific heat capacities are listed on page 136 of this book.

Sample Problem 1
How much heat, in joules, does it take to melt 450 grams of ice initially at –3.0°C to liquid water at 0°C?

Focus: $Q = ?$ We need to add enough heat to first warm the ice from –3.0°C to 0°C ($c_{ice}m_{ice}\Delta T_{ice}$) plus enough heat to melt the ice ($m_{ice}L_{fus}$). So

$$Q = c_{ice}m_{ice}\Delta T_{ice} + m_{ice}L_{fus} = \left(2.09\frac{J}{g\cdot C°}\right)(450\,g)(3.0°C) + 450\,g\left(335\frac{J}{g}\right) = 1.54\times10^5\ J = 154\ kJ.$$

Sample Problem 2
**• An 18-g ice cube at - 4.0°C is placed into 75 g of water at 10°C in an insulated container.
(a) What is the final temperature of the system?**

Focus: $T_f = ?$ Energy flows from the water and warms the ice cube. The water will cool and the ice cube will start to melt.

If the ice cube is very small and the water is very warm, all of the ice will melt and the ice water that forms then warms. The end result is the water becoming a little cooler than it was initially. But if the ice cube is very large, or very cold, or the water we place it in is not very warm, the "warm" water will cool to 0°C before the ice even reaches its melting point. If the ice is still below 0°C it will take heat from the liquid water at 0°C. The water will freeze onto the ice cube even as the ice is still warming.

We need to determine where in this spread of possible outcomes our particular system lies. Let's consider the energy needed to warm up the ice cube first:

Quantity of heat to warm the ice cube to 0°C
$$= c_{ice}m_{ice}\Delta T_{ice} = c_{ice}m_{ice}(T_f - T_0) = \left(2.09\frac{J}{g\cdot°C}\right)(18.0\,g)(4.0°C) = 150\ J.$$
Quantity of heat the liquid water can release as it cools to 0°C
$$= c_w m_w \Delta T_w = c_w m_w (T_f - T_0) = \left(4.18\frac{J}{g\cdot C°}\right)(75\,g)(0 - 10.0°C) = -3135\ J.$$

So the ice will warm to 0°C before the water reaches 0°C. Since 3135 J > 150 J, the ice will begin to melt.

The quantity of heat needed to melt all of the ice $= m_{ice}L_{fus} = (18.0\,g)\left(335\frac{J}{g}\right) = 6030\ J.$

The water doesn't have enough thermal energy to melt all of the ice before itself cooling to 0°C. So the final temperature of the system is **0°C**.

(b) How much of the ice melts?

Focus: $m_{melted} = ?$ We know the ice cube warms to 0°C, then some of it will melt.

Quantity of heat to warm the ice cube + quantity of heat to melt some of ice = −Energy change of the liquid water:

$$c_{ice} m_{ice} \Delta T_{ice} + m_{melted} L_{fus} = -c_w m_w \Delta T_w$$

$$\Rightarrow m_{melted} = \frac{-c_w m_w \Delta T_w - c_{ice} m_{ice} \Delta T_{ice}}{L_{fus}} = \frac{-4.18 \frac{J}{g \cdot C^\circ}(75g)(0 - 10.0°C) - 2.09 \frac{J}{g \cdot C^\circ}(18g)(-4 - 10.0°C)}{335 \frac{J}{g}}$$

$$= \frac{3135\ J - 527\ J}{335 \frac{J}{g}} = 7.8\ \mathbf{g}.\ \text{So about 43\% of the ice melts.}$$

Sample Problem 3

You place a 300-W travel immersion heater into a glass cup that holds 245 grams of water at 24°C. The heat entering the cup is negligible compared to the heat entering the water.

(a) What is the temperature of the water in the cup after 180 seconds?

Focus: $T_f = ?$ Heat will flow into the water at a rate of 300 J/s for 180 seconds, which will raise the temperature of the water. But if water temperature reaches 100°C during the 180-second interval, all of the remaining energy will change the phase of water into steam. Recall that energy = power × time, so $Pt = Q = c_w m_w \Delta T_w$ (+ possibly $m_{steam} L_{vap}$?) .

First let's assume that no water boils, and find the final temperature of the water.

$$Q = Pt = (c_w m_w)(T_f - T_0) \quad \Rightarrow T_f = T_0 + \frac{Pt}{c_w m_w}$$

$$= 24.0°C + \frac{(300 \frac{J}{s})\ 180\,s}{4.18 \frac{J}{g \cdot C^\circ}(245g)} = 24.0°C + 52.7°C = \mathbf{76.7°C}.$$

The resulting temperature of the water and cup is 76.7°C. No energy goes to making steam.

(b) What is the temperature of the water in the cup after 360 seconds?

Analysis: Careful here! Another 180 seconds would increase the temperature of the water by an additional 52.7°C, which would produce a final temperature 76.7°C + 52.7°C = 129.4°C, an impossibility at normal atmospheric pressure. When the water reaches its boiling point, 100°C, continued heat input changes the phase of the water instead of increasing its temperature. So the final temperature of the water in the cup is **100°C**.

(c) How much water remains in the cup after 360 seconds?

Focus: $m_f = ?$ The mass of water left will be the difference between the initial mass and the mass that boiled away.

From $Q_{from\ heater} = Q_{warm\ the\ water\ from\ 24°C\ to\ 100°C} + Q_{boil\ some\ of\ the\ water}$

$$\Rightarrow Q = Pt = (c_w m_w)(T_f - T_0) + m_{vap} L_{vap} \quad \Rightarrow m_{vap} = \frac{Pt - (c_w m_w)(T_f - T_0)}{L_{vap}}$$

$$= \frac{(300 \frac{J}{s})360\,s - \left[(4.18 \frac{J}{g \cdot C^\circ})(245g)\right](100°C - 24.5°C)}{2260 \frac{J}{g}} = 14\,g.$$

The mass of water remaining in the cup is about 245 g − 14 g = **231 g**.

Problems

17-1. Phase changes involve energy changes.
 (a) What quantity of heat must be removed from m grams of liquid water at $0°$ C to turn it into ice at $0°C$?
 (b) Calculate the quantity of heat that must be removed from 45 gram of liquid water at $0°C$ to turn it into ice at $0°C$.

17-2. Energy is needed to melt ice.
 (a) Consider removing m grams of ice from the freezer at a temperature T_0. Later the ice is a puddle of water at T_f. What quantity of heat did the ice absorb?
 (b) A 22-g chunk of ice is removed from the freezer at $-5°C$. Later it is a puddle of water at $17°C$. What quantity of heat did the ice absorb?

17-3. Energy is needed to change the phase of water to steam.
 (a) What quantity of heat must be added to m grams of liquid water at T_0 to turn it into steam at $100°C$?
 (b) Calculate the quantity of heat that must be added to 133 grams of water at $32°C$ to turn it into steam at $100°$.

17-4. Freezing soup is like freezing water. Assume the same thermodynamic parameters for soup as for water.
 (a) What quantity of heat has to be removed from m grams of soup at temperature T_0 to freeze it into frozen soup at T_f $(T_f < 0)$?
 (b) Calculate the quantity of heat that has to be removed from 565 grams of soup at $78°C$ to freeze it into frozen soup at $-5°C$.

17-5. You decide to try the "South-Pole Diet," which consists of eating ice and burning calories to melt the ice. You eat shaved ice at $0°C$ and your body melts the ice and warms it to body temperature, $37°C$.
 (a) How much ice do you have to consume to "burn" 3500 Calories (roughly the energy stored in a pound of fat)? (1 Cal = 1kcal = 1000 cal = 4.184 kJ.)
 (b) Why would you recommend or not recommend the "South-Pole Diet" to overweight friends?

17-6. An M-gram chunk of hot iron is placed on a large piece of $0°C$ ice, whereupon m grams of ice melt.
 (a) What will be the final temperature of the system?
 (b) What must have been the initial temperature of the hot iron?
 (c) A 250-gram chunk of hot iron is placed on the top of a large piece of $0°C$ ice, whereupon 48 grams of ice melt. Calculate the initial temperature of the iron. (The specific heat of iron is 0.44 J/g·$°C$).

17-7. M grams of ice at a temperature T_c are placed in an insulated container holding a large quantity of liquid water at $0°C$.
 (a) How much of the liquid water freezes onto the ice?
 (b) Calculate how much ice is formed on a 25-g piece of ice initially at $-12°C$.

17-8. A mass m_0 of water sits boiling in a pot. After time t the mass of water left in the pot is m_f.
   (a) At what rate is energy being transferred into the pot?
   (b) Calculate the rate of energy transfer if 763 grams of water were initially in the pot and 12.0 minutes later 696 grams of water remain in the pot.

17-9. One way to measure the heat of fusion of ice is to place a known mass of ice at 0°C in some water whose temperature and mass have been measured, let the ice melt, and then measure the final temperature.
   The following lab data:

Mass of water	125 grams
Initial temperature of water	26.0°C
Mass of ice	15.0 grams
Final temperature of the water in the cup	14.6°C

   (a) Calculate the heat of fusion of ice from the data in the table, assuming that no significant amount of energy goes into the Styrofoam cup holding the water.

17-10. One way to measure the heat of vaporization of water in the lab is to boil water in a container that is sealed except for a hose coming out that allows steam to escape. The steam is bubbled into some water in an insulated cup and condenses to form additional liquid water in the cup. The following lab data:

Original mass of water	139.0 grams
Original temperature of water	11.2°C
Mass of water after the steam bubbles in	142.1 grams
Final temperature of the water in the cup	24.8°C

   (a) Calculate the heat of vaporization of water from the data in the table, assuming that no significant amount of energy goes into the Styrofoam cup holding the water. (Hint: Don't forget the cooling of the steam as well as the heating of the original water.)

17-11. One way to cool a swimming pool is to pump water out of the pool and shoot a "spray" of water back into the pool. As the water travels through the air some of it evaporates.
   (a) If the temperature of the water leaving the pool is T_0 and the temperature of the water returning into the pool is T_f, what fraction of the water evaporates while it travels through the air?
   (b) The water returning into the pool is 4°C cooler than the water leaving the pool. What percentage of the water evaporated while in the air? (Note that water can evaporate at ordinary temperature. It doesn't have to boil.)

17-12. Air conditioners are sometimes rated in "tons" of cooling capacity. A ton here is defined as the amount of cooling provided when a ton (2000 lb) of ice is melted over a period of 24 hours.
  (a) How many joules does it take to melt a ton of 0°C ice?
  (b) How many watts is equivalent to a "2-ton" air conditioner?

17-13. A six-pack of canned soda contains m grams of liquid. Suppose that you wanted to place ice into a perfectly insulating ice chest to cool the cans down from T_s to a final temperature T_f. The ice is initially at a temperature T_i.
  (a) How much ice would you need? (Think: The final state of the ice will be as liquid water.)
  (b) Calculate the mass of ice, initially at -5°C, needed to cool 2.1 kg of canned soda from 23°C to a final temperature of 4°C.

17-14. You are lying on the beach on a hot day, neither gaining nor losing energy through conduction with the air. Energy gained from sunshine is counteracted by energy given off by evaporation of perspiration.
  (a) Suppose the Sun shines directly upon your 0.5 m² of exposed skin with an intensity of 700 W/m², and that you can cool your body only through evaporation (that is, sweating). How much water must you sweat each hour to maintain a constant body temperature?

17-15. "Swamp coolers" are sometimes used in dry climates as a form of air conditioning. Hot dry air is drawn through damp cloth or damp wood shavings before entering a building. Some of the heat in the air is used to vaporize the water, thus cooling and humidifying the air at the same time. The density of air is approximately 1.2 kg/m³. Assume that the specific heat capacity of air is about 1000 J/kg·C°.
  (a) To cool a volume V of air (in cubic meters) by an amount ΔT every minute, how many kg of water must evaporate each hour?
  (b) To cool 8.5 m³ of air per minute by 7°C, how much water should evaporate each hour?

17-16. You're about to head out the door with some coffee, which you've poured from the pot into a Styrofoam cup. But the coffee temperature T_h is hotter than you'd like it to be.
  (a) To lower the coffee's temperature to T_c how many grams of ice at 0°C must you add to it? (Remember that you have to warm up the resulting "ice water" to T_c.)
  (b) Calculate how much ice at 0°C should be added to 275 g of coffee at 83°C to lower the coffee's temperature to 71°C.

17-17. You, mass M, get onto your sled of mass m (in kg) at the top of a slope. You slide down the hill of height h and come to rest a short distance past the bottom of the hill on the flat area below.
  (a) Assuming that all of your energy went into melting snow and that the temperature of the snow, both before and after melting, is 0 °C, how much snow did you melt?
  (b) Calculate the mass of snow you melted if your mass is 65 kg and the mass of your sled is 4kg, for a 140-m high hill.

Show-That Problems

17-18. Mixing boiling water with cooler water raises the temperature of the cooler water.
Show that mixing 15.0 L of boiling water to 27.5 L of 15°C water will result in a water temperature of 45°C (assuming no losses).

17-19. Energy is needed to change water to steam.
Show that 217 kJ of energy are required to turn 86 grams of water at 36°C to steam.

17-20. After steam from a steam kettle enters a room and condenses, the resulting water then cools to 22°C, room temperature.
Show that each gram of steam gives up about 2600 J to warm the room.

17-21. When asked how much energy is needed to change the phase of 1 g of ice at 0°C to 100°C steam, a common answer is 2595 J (or 620 calories).
Show that the correct answer is 3013 J (or 720 calories). (What error was made for the incorrect answer?)

17-22. Placing ice in water cools the water.
Show that putting 12.0 grams of ice at 0°C into 82.0 g of water at 23.9°C will produce a final temperature of the system of 10.6°C.

17-23. Hot metal can melt ice.
Show that if 57.3 grams of metal at 100°C is placed into a large chunk of ice at 0°C and 7.5 grams of the ice melts, the specific heat capacity of the metal is 0.44 J/g·C°.

17-24. Aluminum has a specific heat capacity of 0.900 J/g·C°.
Show that when a 10.0-gram piece of aluminum at 20°C gains 225 J of heat that its final temperature will be 45°C.

17-25. An ice cube at 0°C is placed in a cup of 100°C water.
Show that if the ice cube has one-eighth the mass of the 100°C water the final temperature of the water will be 80°C (assuming negligible energy transfer to the cup).

17-26. Equal masses of ice at 0°C and steam at 100°C are added together.
Show that the temperature of the resulting mixture is 100°C.

17-27. When certain amounts of 0°C ice and 100°C steam are added together, the result can be 100°C water.
Show that this will occur when the mass of the ice is three times the mass of the steam.

17-28. When certain amounts of 0°C ice and 100°C steam are added together, the result can be water at 50°C.
Show that this will occur when the mass of the ice is about 4.5 times the mass of the steam.

17-29. Heat is added to 2.00 kg of ice at 0°C to melt it, then to bring it up to the boiling temperature, 100°C, then to turn it to 100°C steam.

Show that the total quantity of heat required is 6.0×10^6 J.

17-30. You bubble 45.9 grams of steam at 100°C into an insulated 113-g copper bucket ($c = 0.386$ J/g·C°) containing 325 g of water and 141 g of ice.

Show that the final temperature of the system will be close to 34.7°C.

18   *Thermodynamics*

Thermodynamics involves the movement of energy from one system to another, and the relationship between heat and work.

Internal Energy

We define U to be the *internal energy* of a system. We can think of internal energy as having two parts:

1. The sum total of the kinetic energies of the individual particles or molecules within the system, in all of its different forms—the linear (*translational*) motion of the particles (related to the temperature of the system) as well as the internal vibrations and rotations of the particles themselves.
2. The sum total of all of the potential energy stored in the attractions and repulsions among all of the particles in relation to one another.

The internal energy of the system can be increased by adding heat to the system, or by doing work *on* the system. If you add heat Q to the system its internal energy will increase—the particles within it will move faster and the temperature will increase, or perhaps the added heat will cause a phase change and increase the potential energy of the system. Likewise, if you do work W on the system (say, on a gas by compressing it—applying a force for a distance on the piston) its internal energy will again increase.

In sum, $\Delta U = Q + W$.

If heat flows out of the system, Q is negative. Likewise, if you let the system do work on the environment (say, an expanding gas that supplies a force for some distance on the environment) the sign of W is negative.

A physicists' favorite system when thinking about thermodynamics is a gas trapped in a piston with a movable cylinder. Heat flowing into the system will raise the temperature of the gas, as will doing work on the system by compressing the gas. Heat flowing out of the system will lower the temperature of the gas, as will letting the gas expand and do work on the environment.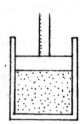

Heat Engines

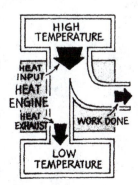

A *heat engine* is any device that converts heat to work in a cyclic fashion. For example, an automobile engine is a heat engine. Burning fuel adds heat to the cylinders. The expanding combustion products do work to accelerate the car or push it against friction. As each cylinder is compressed to get ready for the next burst of input heat, a smaller amount of heat is exhausted from the engine. The diagram at the left shows an idealized view of a heat engine. At some high temperature, heat Q_{in} is added to the engine, which then does work W on its surroundings and ejects heat Q_{out} at a lower temperature. By energy conservation, $W = Q_{in} - Q_{out}$.

The *efficiency* of a heat engine $\varepsilon = \dfrac{\text{Work out of the system}}{\text{Heat into the system}} = \dfrac{W}{Q_{in}} = \dfrac{Q_{in} - Q_{out}}{Q_{in}}$.

The French engineer Sadi Carnot determined that there was a limit to how efficient a thermodynamic cycle could be—you can never turn all of the heat input into the system into work. Energy has to flow into the heat engine at a high temperature T_{hot}, and some energy has to flow out to a cooler environment at T_{cold}. The best efficiency *ever* for a heat engine operated between a high-temperature reservoir and a low temperature reservoir is given by:

Ideal efficiency $\varepsilon = \dfrac{T_{hot} - T_{cold}}{T_{hot}} = 1 - \dfrac{T_{cold}}{T_{hot}}$ (with temperatures in *kelvins*).

Carnot's calculations assume that the cycle happens in a series of tiny incremental steps, as though the piston was being moved slowly, one micron at a time, with plenty of time for the system to reach equilibrium before the next one micron movement of the piston. Real heat engines do work in a finite amount of time and their efficiency is always lower than that of a Carnot heat engine.

Adiabatic Processes in the Atmosphere

Compressing or expanding a gas while no heat enters or leaves the system is said to be an adiabatic process. As a parcel of warm air rises through the atmosphere the pressure around it decreases and the parcel of air expands. The parcel does work on the surrounding air as it expands against the external pressure. This work comes at the expense of the internal energy of the air mass, so its temperature decreases—about 10°C for each kilometer the air rises in the atmosphere. Likewise, as a parcel of cool air descends the increasing pressure of the atmosphere compresses it and makes it hotter (also 10°C per km).

Sample Problem 1

Q joules of heat are added to a gas while *W* joules of work are done in compressing it.

(a) Calculate the change in the internal energy of the gas.

Focus: $\Delta U = ?$
The internal energy change in general depends on both the amount of heat flowing into or out of the system, and the amount of work done on or by the system. In this case heat flows into the gas while work is done on it. So $\Delta U = Q + W$.

(b) Calculate the change in the internal energy of the gas when 100 J of heat are put into it while at the same time 60 J of work are done in compressing the gas.

Solution: $\Delta U = Q + W = (+100 \text{ J}) + (+60 \text{ J}) = \textbf{+160 J}$.

Sample Problem 2
A sample of gas absorbs heat while it expands and does work on its surroundings.

(a) Calculate the change in the internal energy of the gas.

Solution: As before, $\Delta U = Q + W$.

(b) Calculate the change in internal energy if the sample absorbs 120 J of heat while expanding and doing 80 J of work on its surroundings.

Solution: $\Delta U = Q + W$. In this case $Q = +120$ J (heat is being added to the gas) while $W = -80$ J (since work is being done *by* the gas instead of *on* the gas). So $\Delta U = Q + W = (+120 \text{ J}) + (-80 \text{ J}) = +40$ J.

Sample Problem 3
In the 1990s a 210-kW OTEC (Ocean Thermal Energy Conversion) power plant in Hawaii was designed to run between the warm ocean surface waters at T_{hot} and the colder deep ocean waters at T_{cold}.

(a) What is the theoretical maximum efficiency of this power plant?

Answer: $\varepsilon = \dfrac{T_{hot} - T_{cold}}{T_{hot}}$ with temperatures expressed in kelvins.

(b) Theoretically, how much heat must be extracted from the warmer seawater each second to produce 210 kW?

Focus: $Q_{in} = ?$
210 kW = 210 kJ of energy *out* each second. This useful work out of the power plant is only a fraction ε of the total heat taken in by the plant.

From $\varepsilon = \dfrac{W_{out}}{Q_{in}} \Rightarrow Q_{in} = \dfrac{W_{out}}{\varepsilon} = \dfrac{210 \text{ kJ}}{\varepsilon}$.

(c) Calculate answers for (a) and (b) when $T_h = 26°C$ and $T_c = 6°C$.

First we must express the given temperatures in Kelvins. They are $T_{hot} = 273 + 26 = 299$ K and $T_{cold} = 273 + 6 = 279$ K. Then

$$\varepsilon = \frac{T_h - T_c}{T_h} = \frac{299\text{K} - 279\text{K}}{299\text{K}} = 0.067.$$ Ideally, 6.7 joules out of every 100 joules of heat taken in by the plant would be turned into electrical energy.

$$Q_{in} = \frac{210\text{kJ}}{\varepsilon} = \frac{210\text{kJ}}{0.067} = 3100 \text{ kJ each second.}$$

Since real power plants do not operate at ideal efficiencies, we'd have to take in more than 3100 kJ per second to produce 210 kW of power.

Sample Problem 4

Glider pilots usually don't carry heaters in their gliders. As altitude is gained the air temperature drops according to the 10-degrees-per-kilometer rule (described on page 347 in the textbook). So glider pilots should dress warmly even though the temperature at the gliderport is comfortably warm.

(a) **In taking off from a gliderport where the temperature is a comfortable T_0, what temperatures can a pilot expect when soaring upward a distance Δh above the ground?**

Focus: $T = ?$

As the glider rises we can assume that the temperature in the cabin is much the same as the air around it, which decreases approximately 10°C for each 1 km (1000 m) increase in elevation. So $T = T_0 - \frac{10°C}{1000\,m}\Delta h$.

(b) **The temperature at the gliderport is 22°C. Calculate the temperature at an altitude 3000 m above the gliderport.**

Solution: $T = T_0 - \frac{10°C}{1000\,m}\Delta h = 22°C - \left(\frac{10°C}{1000\,m}\right)3000\,m = -8°C$.

Problems

18-1. In each case below calculate the missing quantity in the table:

	Q	*W*	*ΔU*
a.	100 J of heat flows in	System does 40 J of work	?
b.	200 J of heat is added	No work is done	?
c.	500 J of heat is added	System does 350 J of work	?
d.	70 J of heat is removed from a gas	70 J of work is done to compress the gas	?
e.	?	Expanding gas does 500 J of work	Internal energy increases by 340 J
f.	You lose 9.0×10^5 J of heat playing basketball	?	Your internal energy decreases by 1.20×10^6 J
g.	No heat is added	40 J of work done to compress a gas	?

18-2. An automobile engine runs at T_h (°C) on a day when the temperature outside is T_c (°C).
(a) What is the maximum possible efficiency of the automobile engine?
(b) Calculate the maximum possible efficiency if the engine runs at 630°C on a 25°C day.

18-3. An ideal heat engine takes in heat Q_{in} at a temperature T_h. It exhausts heat Q_{out}.
   (a) How much work is done by the engine?
   (b) What is the efficiency of the engine?
   (c) What is the exhaust temperature of the engine?
   (d) Calculate answers to the above for a heat input of 460 J at a temperature of 600 K, and a heat output of 285 J.

18-4. A heat engine exhausts heat Q_{out} while performing useful work W.
   (a) What is the efficiency of the engine?
   (b) Calculate the efficiency if the engine does 1200 J of work while exhausting 3800 J of waste heat to the environment.

18-5. A heat engine takes in heat Q_{in} and exhausts heat Q_{out}.
   (a) What is the efficiency of the engine?
   (b) Calculate the efficiency of an engine that takes in 3600 J of heat and exhausts 1400 J of heat.

18-6. A particular heat engine has efficiency ε.
   (a) How much heat must it take in to do work W?
   (b) How much heat will it exhaust to its surroundings?
   (c) Calculate answers for (a) and (b) for a 27% efficient heat engine that does 1400 J of work.

18-7. Dr. Knute C. Cuckoo claims to have invented a heat engine that will revolutionize "life as we know it." It runs between a hot source at 300°C and cold heat "sink" at 25°C. Dr. C. claims that his engine is 92% efficient.
   (a) What is the actual maximum efficiency of his heat engine?
   (b) What error did he make in his calculations?

18-8. A geothermal power plant runs between hot groundwater at T_h and a river at T_c.
   (a) What is the theoretical maximum efficiency of this power plant?
   (b) Calculate the maximum efficiency for $T_h = 185°C$ and $T_c = 15°C$.

18-9. A geothermal power plant runs between underground hot water at T_h and a river at T_c.
   (a) If the plant produces x kW of power, at least how much heat must it absorb from the hot groundwater every second?
   (b) Calculate the minimum amount of heat that must be absorbed each second by a 35- kW geothermal power plant that operates between $T_h = 185°C$ and $T_c = 15°C$.

18-10. A Stirling engine is a kind of heat engine that typically uses two joined piston-cylinder combinations to obtain work from an external source of heat. Consider a Stirling engine that heats a confined gas to T_h and rejects heat to its surrounding at T_c. It has a solar collector of area A (m^2), which is illuminated by sunlight of intensity I (W/m^2).
(a) What is the theoretical maximum efficiency of this Stirling engine?
(b) How much sunlight energy hits the collector each second? (Answer will be in J/s = watts.)
(c) In theory, how much useful work can we get out of this Stirling engine each second?
(d) In practice, the engine runs at 75% of its theoretical efficiency. How much useful work per second do we actually get out of the engine?
(e) This Stirling engine has a solar collector area of 3.0 m^2 and operates with a high temperature of 300°C and its surroundings are at 30°C. Calculate the actual power produced by the engine if the sunlight intensity is 800 W/m^2.
(f) Another Stirling engine using the same solar collector operates with a high temperature of 100°C in 30°C surroundings. It runs at 70% of its ideal efficiency. Calculate the actual power produced by this lower temperature engine if the sunlight intensity is 800 W/m^2.

18-11. A small geothermal electrical generation site Mladost-Zagreb in Croatia uses hot water from the ground as the high temperature reservoir, and the outside air as the low temperature reservoir. Hot groundwater enters the plant at a temperature T_0 and leaves the plant at a slightly lower temperature T_f. Waste heat is rejected to the environment at a temperature T_c. Hot groundwater of m kg is pumped through the power plant each second.
(a) How much heat is supplied to the power plant every second? (The hot groundwater is cooled from T_0 to T_f.)
(b) What is the theoretical maximum efficiency of the geothermal plant, if the temperature of the hot "reservoir" is taken to be the average of T_0 and T_f? What fraction of the heat input can this heat engine ideally turn into work?
(c) What is the maximum power output of the geothermal plant?
(d) At the geothermal plant in Mladost-Zagreb 5.8 kg per second of hot groundwater enters the plant at $T_0 = 79.6$°C and leaves the plant at 79.4°C. In the winter, the temperature of the outside air is $T_c = -0.4$°C. Calculate the theoretical power output of this geothermal plant.

18-12. Turbine generators are like jet engines—they draw in large quantities of air, then compress it and mix it with fuel. The combustion of the fuel turns the turbine, which draws in and compresses more air, and also turns a generator. Typical combustion temperatures are around 1100°C and typical exhaust temperatures are around 500°C.

(a) If this turbine operates like an ideal Carnot heat engine, what would be the efficiency of the turbine?

(b) For every 100.0 J of heat the engine takes in, how much work is done?

(c) For every 100.0 J of heat the engine takes in, how much heat is rejected into the exhaust?

(d) *Combined-cycle* turbines take the hot exhaust and use it as a source of heat to produce steam to run a steam turbine. If the exhaust from the gas turbine (now our heat *input*) is at 500°C, and the cooling water for the steam turbine is at 20°C, what is the maximum efficiency of this second part of the cycle?

(e) In part (c) you calculated the amount of waste heat from the first part of the cycle, which is also the heat input for the second part of the cycle. Use the efficiency that you calculated in part (d) to calculate how much work you get out of the steam turbine.

(f) How much work is done overall in the *combined*-cycle turbine for every 100.0 J of heat input? [This should be the work you calculated in (b) *plus* the work you calculated in (e)].

(g) What is the theoretical efficiency of the combined-cycle turbine?[*]

(h) Why does using a combined-cycle turbine make more sense than using a single-cycle turbine?

18-13. A mass of air descends from a height h_0 above sea level to a height h_f above sea level.

(a) What is the change in temperature of the air?

(b) Calculate the change in temperature for dry air that descends from 6200 m above sea level to 2500 m above sea level.

Show-That Thermodynamics Problems

18-14. In a certain operation, 160 J of work is done on a gas while 50 J of heat are removed from the gas.
Show that its change in internal energy is +110 J.

18-15. A gas does 250 J of work while absorbing 130 J of heat.
Show that its change in internal energy is -120 J.

18-16. A certain system experiences an internal energy change of +670 J while absorbing 1350 J of heat.
Show that the system does 680 J of work on its surroundings.

18-17. A certain heat engine takes in 25 kJ of heat and exhausts 17 kJ.
Show that its efficiency is 0.32.

18-18. A heat engine operates between $T_{hot} = 750°C$ and $T_{cold} = 35°C$.
Show that the theoretical maximum efficiency is about 0.70, or 70%.

18-19. A power plant has a theoretical maximum efficiency of 0.46 and an exhaust temperature of 35°C.
Show that the high-temperature reservoir must be at 297°C.

18-20. A 420 kW power plant runs between 540°C and 30°C.
Show that the plant requires a heat input of at least 670 kJ per second.

18-21. Air rises from an elevation of 800 m above sea level, where the temperature is 23°C, to an elevation of 4500 m above sea level.
Show that when this occurs the air's temperature likely drops to -14°C.

18-22. The 86th floor observation deck of the Empire State Building is 320 m above the ground.
Show that the temperature there is normally 3.2°C cooler than the temperature at ground level.

19 Vibrations and Waves

Many things vibrate or oscillate—a pendulum, an object on the end of a spring, a tuning fork, the strings of musical instruments, or the air columns in wind instruments. Oscillations can produce waves, and waves can be either *standing* (as in the vibration of a violin string) or *traveling* (as in the sound wave that carries energy through the air from the violin to your ear). Waves are characterized by an amplitude A, a wavelength λ, a frequency f, and a speed v. A very useful relationship is $v = \lambda f$. That is, a wave's speed is equal to the product of its wavelength and frequency.

Sample Problem 1
A sine wave has amplitude A and wavelength λ.

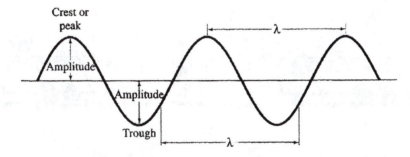

(a) What is the horizontal distance between the crest of a wave and the nearest trough?

Answer: Inspection of the figure shows this distance to be one-half a wavelength: $\lambda/2$.

(b) If the speed of the wave is v, what is its frequency?

Focus: $f = ?$ From $v = f\lambda \Rightarrow f = \dfrac{v}{\lambda}$.

(c) Calculate the frequency if the wave speed is 1.5 m/s and the wavelength is 2.5 m.

Solution: $f = \dfrac{v}{\lambda} = \dfrac{1.5 \frac{m}{s}}{2.5\,m} = \mathbf{0.60\,m}$.

Sample Problem 2
While sitting at the dock of the bay, Otis notices incoming waves with distance d between crests. The incoming crests lap against the pier pilings at a rate of one every 2 seconds.
(a) Find the frequency of the waves.

Answer: The frequency of the waves is given, one per 2 s, or $f = \mathbf{0.5\ Hz}$.

(b) Find the speed of the waves.

   Solution: $v = f\lambda = fd$.

(c) Calculate the speed of the waves if the distance d between crests is 1.8 m.

   Solution : $v = fd = 0.5\,\text{Hz}(1.8\,\text{m}) = 0.5\left(\frac{1}{s}\right)(1.8\,\text{m}) = \mathbf{0.9\,\frac{m}{s}}$.

The Doppler Effect

The pitch of a sound depends on its frequency, given by $f = v/\lambda$, where v and λ are the speed and wavelength of sound, respectively. When a car honking its horn approaches you, the compressions of the sound waves arrive at your ear closer together and with a frequency that is higher than it would be if the car were at rest. You hear a higher pitch than the horn is producing. This is the Doppler effect. When a car recedes from you, the compressions of the sound waves arrive at your ear farther apart and with a lower frequency. Now the pitch you hear is lower—again, the Doppler effect. The frequencies heard by the observer, f_o, are related to compressed or stretched wavelengths, and are given by these equations:

$$f_o = \frac{v}{v - v_s}f_s \qquad\qquad\qquad f_o = \frac{v}{v + v_s}f_s$$

Frequency heard for approaching source   Frequency heard for receding source

The frequency of the source is f_s, the frequency heard by the observer is f_o, the speed of the source is v_s and speed of sound is v.[*]

Sample Problem 3

A fire engine moving at speed v_s with its siren blasting away approaches you and soon thereafter recedes. The frequency of the siren is f_s.

(a) Calculate the frequency you hear when the vehicle approaches at a speed of 30 m/s with its siren blasting at 500 Hz. (The speed of sound in air is 334 m/s.)

   Focus: f_o = ? Which equation do you use? Since the siren is approaching you, you expect to hear a higher frequency. Both expressions for the observed frequency are of the form f_o = ratio$\times f_s$. You need the ratio that is greater than 1, where you divide v by a number less than v. You choose the expression with $v - v_s$ in the denominator.

   Solution: $f_o = \dfrac{v}{v - v_s}f_s = \dfrac{334\ \text{m/s}}{334\ \text{m/s} - 30\ \text{m/s}}500\ \text{Hz} = \mathbf{550\ Hz}$.

_____

[*] Derivations of these formulas, and formulas for cases where the observer is in motion as well, can be found in many algebra-based physics textbooks. We won't treat these extensions of the Doppler effect here.

(b) Calculate the frequency you hear when the vehicle recedes at the same speed.

Focus: $f_o = ?$ Since the siren is receding, you will hear a lower frequency. You need the ratio in the equation to be less than 1, so you choose the expression with $v + v_s$ in the denominator.

Solution: $f_o = \dfrac{v}{v + v_s} f_s = \dfrac{334 \text{ m/s}}{334 \text{ m/s} + 30 \text{ m/s}} \, 500 \text{ Hz} = \mathbf{460 \text{ Hz}}.$

(c) Why do firemen on the moving vehicle hear no Doppler shift in the siren?

Answer: They hear no Doppler shift because there is no velocity difference between them and the siren. The Doppler effect results from *differences* in velocities between the source and the receiver.

Shock Wave

A supersonic (faster-than-sound) aircraft produces a conical shock wave composed of overlapping spheres of sound crests, similar to the bow wave of a boat composed of overlapping circular waves. We see in the figure that while the aircraft moves a distance $v_s t$, sound moves a distance vt (where v_s is the speed of the source and v is the speed of sound). Note that $\sin\alpha = \dfrac{vt}{v_s t} = \dfrac{v}{v_s}$. The faster the aircraft, the smaller is $\sin\alpha$ and the narrower the cone. Interestingly, the inverse of this ratio, v_s/v, is called the Mach number. For example, at twice the speed of sound, $v_s/v = 2$ and we say the craft is moving at Mach 2. The angle α shown in the sketch is $\sin^{-1}\left(\frac{1}{2}\right) = 30°.$ This is half the angle of the V shape produced by the aircraft, which is 60°.

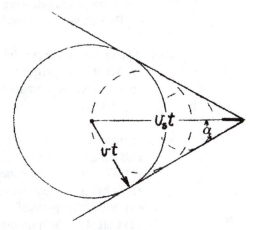

Sample Problem 4

When the Concorde recently flew at its cruising speed of Mach 1.75 it created a sonic boom.
(a) How fast in km/h was its cruising speed? (Use 340 m/s for the speed of sound in air.)

Solution: Cruising speed was $1.75\left(340\,\frac{m}{s}\right) = 595\,\frac{m}{s}$. Converted to km/h,

$$595\frac{m}{s} \times \frac{1 \text{ km}}{1000 \text{ m}} \times \frac{3600 \text{ s}}{1 \text{ h}} = 2140\frac{\text{km}}{\text{h}}.$$

(b) From the moment the Concord is 8 km directly overhead, how many seconds elapse before you hear its sonic boom?

Focus: $t = ?$ You can imagine that the nose of the sonic boom cone starts forming when the Concorde is directly above you. We're looking for the time for the sound waves to travel to you, 8 km below.

From $v = \dfrac{d}{t} \Rightarrow t = \dfrac{d}{v} = \dfrac{8000 \text{ m}}{340\frac{m}{s}} = \mathbf{24 \text{ s}}.$

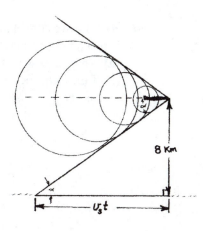

(c) How far past the overhead point would the Concorde be when the boom reached you?

Solution : $d = vt = 595 \frac{m}{s} (24 \, s) = 14,000 \, m = \textbf{14 km}.$

At high altitudes, where it is cold, the speed of sound is somewhat less than 340 m/s, so in fact the Concorde would be even farther ahead.

Problems

19-1. A wave travels along a stretched rubber tube. The vertical distance from crest to trough for the wave is Y and the horizontal distance from crest to nearest trough is X.
(a) What is the amplitude of the wave?
(b) What is the wavelength of the wave?

19-2. A simple pendulum swings to and fro once in time t.
(a) What is the frequency of the pendulum?
(b) If the pendulum were shorter, would the frequency be more or less? Defend your answer.

19-3. The period of the pendulum in a grandfather clock is T.
(a) What is the frequency of oscillation?
(b) If its period were longer, would its frequency be greater or less? Defend your answer.

19-4. A mass on a vertical spring vibrates over a distance d from its lowest to its highest point and completes one full vibration up and down in time t.
(a) What is the amplitude of vibration?
(b) What is its frequency?
(c) What is its period?
(d) Calculate the amplitude, frequency, and period if d is 0.30 m and t is 2.0 s.

19-5. Light travels at 3.00×10^8 m/s through a vacuum, and for most practical purposes, through air as well.
(a) Find the wavelength of light of frequency 5.09×10^{14} Hz emitted by a sodium lamp.
(b) What is the frequency of light whose wavelength is 5.5×10^{-7} m?

19-6. Radio waves travel at the speed of light, 3.00×10^8 m/s. The wavelength of radio waves is much longer than the wavelength of visible light waves.
(a) What is the wavelength of radio waves having a frequency of 500 kHz?
(b) If the frequency were higher, would the waves be longer or shorter? Defend your answer.

19-7. A mosquito flaps its wings at frequency f, which produces the annoying buzz that travels at the speed of sound, v.
   (a) What is the frequency of the buzz?
   (b) How far does the sound travel between wing beats?
   (c) What is the wavelength of the annoying sound?
   (d) Calculate the wavelength of the buzzing sound if the frequency of vibration is 600 Hz and the speed of sound is 340 m/s.

19-8. While sitting on a pier, Lillian notices water waves incident upon the pier pilings at regular intervals.
   (a) Calculate the speed of the waves if the distance between their crests is 3 m and they hit the pilings 20 times per minute.

19-9. Two buoys float a distance d apart in a channel. They rise and fall once every second.
   (a) Calculate the wavelength of the waves if the speed of the waves is 5.0 m/s.
   (b) Calculate the number of waves between the buoys if they are 200 m apart.

19-10. Steve produces waves in a small pond by dipping his hand into the water at regular intervals. He observes the waves traveling between a pair of reeds that stick out of the water a distance d apart.
   (a) It takes t seconds for the first wave to travel past the first reed to the second reed. What is the speed of the waves?
   (b) Steve dips his fingers into the water at a frequency f. What is the wavelength of the waves?
   (c) At any moment, how many waves are there between the reeds?
   (d) The reeds are 1.5 m apart, and when Steve dips his finger into the water two times each second the waves take 3.0 second to pass between the first reed and the second reed. Calculate the speed and wavelength of the waves, and the number of waves between the two reeds.

19-11. Two canoes tied to a dock are distance d apart. One is at a crest at the same time the other is at an adjacent trough (one-half wave between them). They each bob up and down at frequency f.
   (a) What is the speed of the waves?
   (b) Calculate the speed if the canoes are 6.0 m apart and they bob up and down once every 2.0 s.

19-12. A wave travels at speed v in a vibrating stretched rubber tube of length L as shown. The fundamental frequency of vibration is f.
   (a) How many nodes exist for the tube?
   (b) Find the wavelength of the fundamental.
   (c) This standing wave can be treated as the superposition of two waves traveling in opposite directions. What is the speed of those waves?

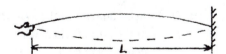

19-13. The frequency of vibration is increased 3-fold for the string of the previous problem.
   (a) How many nodes exist for the tube?
   (b) Find the wavelength of the new standing wave.

19-14. As a fire engine approaches you, its 500-Hz horn sounds and your sound analyzer measures the frequency that you hear.
   (a) If the sound analyzer registers 520 Hz, calculate the speed of the fire engine (assume a 340-m/s speed of sound).
   (b) What would your meter read if the same fire engine were moving away at the same speed?

19-15. While standing near a railroad crossing you hear the horn of an approaching train.
   (a) You hear a frequency of 431 Hz, and the train is moving toward you at 28 m/s (100 km/h). Calculate the horn frequency that the train conductor hears. The speed of sound in air is 340 m/s.
   (b) Why does the train conductor not hear a Doppler shift in the horn's frequency?
   (c) Calculate the frequency you would hear if the train were receding at the same speed.

19-16. A train whistle emits sound waves at a frequency of 500 Hz, but is heard as 550 Hz by a stationary listener. The speed of sound in air is 340 m/s.
   (a) Is the train approaching or receding from the listener?
   (b) Calculate the speed of the train.

19-17. You, at rest, hear the sound of an approaching source as having a higher pitch due to the Doppler effect.
   (a) Calculate the speed of an approaching source that would increase the sound frequency you perceive by 10%.
   (b) If the source were receding at the same speed, by what percent would the frequency be lower?

19-18. Stationary Stan is listening to the sound from an airplane flying along his line of sight. The frequency perceived by Stan is half the frequency emitted by the plane.
   (a) Would the airplane be approaching or receding?
   (b) Calculate the speed of the airplane if the speed of sound in air is 340 m/s.

19-19. A speedboat makes a bow wave as it knifes through the water.
   (a) What is the angle of the resulting V shape of the bow wave if the boat travels 5 times as fast as waves on the surface? (In the diagram in the discussion of the shock wave, this angle is 2α.)
   (b) As the boat gains speed, does the angle of the bow wave increase or decrease?

19-20. A supersonic aircraft produces a conical shock wave in the air.
   (a) If the aircraft flies at 1.5 times the speed of sound, what is the angle (2α) of the conical wave?
   (b) If the aircraft flies faster, does the angle increase or decrease?

Show-That Problems

19-21. A cork floating in water bobs up and down four times each second as a small wave passes. The speed of the wave is 0.40 m/s.
Show that the wave peaks are 0.10 m apart.

19-22. A mechanical metronome can be set to vibrate at a desired frequency.
Show that if it makes 9 complete vibrations in 12 s its frequency is 0.75 Hz.

19-23. A tall skyscraper swings to and fro in strong winds at a frequency of 0.15 Hz.
Show that the period of this vibration is 6.7 s.

19-24. Ultraviolet light has a higher frequency than visible light.
Show that the frequency of ultraviolet light of wavelength 360 nm is 8.33×10^{14} Hz.
(1 nm = 10^{-9} m.)

19-25. A nurse counts 84 heartbeats in one minute.
Show that the period and frequency of the heartbeats are 0.71 s and 1.4 Hz respectively.

19-26. The crests of a long ocean wave are 24 m apart and in one minute 10 crests pass by.
Show that the speed of the wave is 4.0 m/s.

19-27. A typical tidal wave (tsunami) can have a speed of 750 km/h and a wavelength of 320 m.
Show that the period of the wave is about 1.5 s.

19-28. A seismic P wave traveling at 8000 m/s has a wavelength of 2000 m.
Show that the frequency of the wave is 2.5 Hz.

19-29. An earthquake on the opposite side of Earth generates P waves that travel through the Earth at an average speed of 8000 m/s.
Show that the time to make the trip through Earth is about 42 minutes.

19-30. Planet Earth, like a spherical bell, vibrates in a discrete set of fundamental frequencies between 0.002 and 0.007 Hz.
Show that the periods of these vibrations range from 143 s to 500 s.

19-31. Seismologists attribute 0.2-Hz micro-seismic vibrations at Earth's surface to ocean wave interactions.
Show that the period of this small vibration is 5 s.

19-32. While watching water at the dock of the bay, Otis notices that ten waves pass him each 30 seconds, and that the crests of successive waves exactly coincide with posts that are 3.6 m apart.
Show that the speed of the waves is 1.2 m/s.

19-33. A standing wave between fixed supports consists of 3 half-wavelength segments.
Show with a sketch that the number of nodes in the standing wave is 4.

19-34. A standing wave between fixed supports has 3 half-wavelength segments.
Show with a sketch that the number of antinodes in the standing wave is 3.

19-35. While standing at a railroad crossing you hear the whistle of an oncoming train.
Show that you'll hear the 800-Hz whistle as 863 Hz when the train's speed is 25 m/s and the speed of sound in air is 340 m/s.

19-36. A humpback whale produces a 99 Hz sound wave as she swims at 15 m/s towards her stationary mate. The speed of sound in seawater is 1530 m/s.
Show that her mate perceives the sound wave as 100 Hz.

19-37. The apparent frequency of sound from a source approaching you is higher than if the source were stationary.
Show that the apparent frequency increases by 5 % when the source approaches you at 16 m/s. (Take the speed of sound in air to be 340 m/s.)

19-38. Your talent for perfect pitch comes in handy. While waiting at the bus stop you note that the familiar 450-Hz horn of an oncoming bus is 470 Hz.
Show that the bus is approaching you at a speed of approximately 14 m/s if the speed of sound is 340 m/s.

19-39. In an ideal situation, a boat knifing through the water makes a bow wave very similar to the conical shock wave made by supersonic aircraft.
Show that the full angle of the bow wave is about 29° when the boat moves at 4 times the speed of water waves on the surface.

19-40. An airplane traveling at twice the speed of sound makes a conical shock wave.
Show that the angle of the cone from its axis (half angle for the shock wave) is about 30°.

20 Sound

This chapter logically extends the material of the previous chapter, focusing specifically on sound waves. The speed of sound is different in different materials. In air at 0°C and 1 atmosphere of pressure, sound travels at a speed of 331 m/s. The speed of sound in air increases approximately 0.60 m/s for each Celsius degree increase in temperature. In the problems below we'll use an average speed of sound of 340 m/s in air (1224 km/h or 723 mi/h).

Sample Problem 1
Sound travels more than four times faster in seawater than in air.
(a) If the frequency of a sound wave in air is 440 Hz, what is its frequency in seawater?

> *Answer*: The frequency of a sound wave is the same in both air and water: **440 Hz**, in this case. The frequency of the sound wave mirrors the vibrational frequency of its source. If the source object vibrates 440 times per second, the medium will be compressed and expanded at exactly the same rate, 440 Hz, independent of what the medium is, even when a wave travels from one medium to another. Wave properties that *can* change as a wave passes from one medium to another are wavelength and speed.

(b) Calculate the wavelength of a 440-Hz note in air.

> *Focus*: $\lambda = ?$

$$\text{From } v = f\lambda \Rightarrow \lambda = \frac{v}{f} = \frac{340\,\frac{m}{s}}{440\ \text{Hz}} = \frac{340\,\frac{m}{s}}{440\left(\frac{1}{s}\right)} = 0.77\ \text{m} = \textbf{77 cm.}$$

(c) Calculate the wavelength of a 440-Hz sound wave in seawater. The speed of sound in seawater is 1530 m/s.

$$\text{Solution}: \lambda = \frac{v}{f} = \frac{1530\,\frac{m}{s}}{440\ \text{Hz}} = \frac{1530\,\frac{m}{s}}{440\left(\frac{1}{s}\right)} = \textbf{3.48 m.}$$

Sample Problem 2
While hiking you note a flash of lightning, and hear the thunder after a time interval t.
(a) How far are you from the bolt of lighting if the time interval between seeing the flash and hearing the thunder is t ?

> *Focus*: $d = ?$
> The speed of light is so much greater than the speed of sound that we can assume that the light reaches you instantaneously. The sound wave arrives at your ears after a time t.

$$d = v_{\text{sound}}t.$$

(b) How far are you from the bolt if you hear the thunder 3.0 seconds after you see the lightning flash?

Solution: $d = v_{sound}t = 340 \frac{m}{s}(3.0 \text{ s}) = 1000 \text{ m} = 1 \text{ km}.$
Sound travels about 1 kilometer in 3 seconds.

Sample Problem 3

A dolphin emits ultrasonic sound waves to locate prey. The closer the prey, the shorter the time for the echo of the ultrasound from the prey to return to the dolphin.
(a) Calculate the distance between a dolphin and its prey if the time between sending and receiving a pulse is 0.22 s. Assume the speed of sound in the water is 1530 m/s.

Focus: $d = ?$
Solution: $d = vt = 1530 \frac{m}{s}(0.11 \text{ s}) = 170 \text{ m}.$
Notice we use 0.11 s for the time, since the 0.22 s is the round-trip time.

(b) How does a Doppler shift in the echo let the dolphin know whether the prey is approaching or receding?

Answer: If the prey is approaching, the frequency of the echo will be higher; if the prey is receding, the frequency of the echo will be lower. Dolphins employ both the time of echo return and the Doppler effect to better assess their prey and their surroundings.

Problems

(For the following problems assume the speed of sound in air is 340 m/s.)

20-1. A factory whistle emits a loud blast of sound. The intensity of the sound diminishes with distance according to the inverse-square law.
   (a) How far away would a listener detect sound half as intense as when at distance d from the whistle?
   (b) How far away would a listener detect sound one-tenth as intense as at distance d from the whistle?

20-2. A sound wave in air has speed v and frequency f.
   (a) What is its wavelength?
   (b) Calculate the wavelength if the frequency is 880 Hz.

20-3. You watch a carpenter down the street hammering nails in a roof and note that the sound and sight are synchronized perfectly. Then when he stops hammering you hear one more hammer blow.
   (a) If the time delay is 1 second, what is the distance between you and the carpenter?
   (b) If you hear two blows after seeing the last blow, what is the distance between you and the carpenter?

20-4. At a musical concert you sit far back and listen to the same concert broadcast on local radio.
  (a) Which do you hear first, the sounds from the radio or from the concert stage?
  (b) If there is a delay of 0.5 s between the radio signal and the sound from the stage, how far from the stage are you? (Assume the radio signal is practically instantaneous compared with the speed of sound, for the radio signal travels at the speed of light.)
  (c) Pretend that in some way the radio signal traveled all the way around the world before reaching you. How far back from the stage would you have to sit so both the radio signal and the sound from the stage traveling through the air arrive at the same time?

20-5. Sound travels about 4.5 times faster in seawater than in air.
  (a) Calculate the speed of sound in seawater.
  (b) Calculate the wavelength in seawater of a sound wave whose frequency is 1500 Hz.

20-6. Lizzie is listening to music under water in a swimming pool that is equipped with underwater speakers.
  (a) Calculate the wavelength of a 264-Hz note in air.
  (b) Calculate the wavelength of a 264-Hz note in water. Assume the speed of sound in fresh water is 1490 m/s.
  (c) Would the 264-Hz note have a different pitch under water than in air? Why or why not?

20-7. The speed of sound in a railroad track is about 4500 m/s. Tom, some distance away, taps the track with a hammer. Pat, with her ear on the track, hears the sound 0.50 s later.
  (a) How far apart are Tom and Pat?
  (b) How long would it take for the sound of the tap to reach Pat through air?

20-8. At a baseball game you hear the crack of the bat hitting a ball 0.30 s after observing it.
  (a) How far are you from home plate?
  (b) If the wind is blowing in your direction, will the time between seeing and hearing the hit change? If so, in what way? Defend your answer.

20-9. You're watching fireworks in the sky. You note a delay between seeing and hearing the explosions.
  (a) How is the duration of the delay between seeing and hearing the fireworks affected if you are twice as far away?
  (b) How does a three-times-as great distance affect the duration of the time delay?

20-10. A whale emits a sound directed toward the bottom of the ocean, a distance y below. The speed of sound in seawater is 1530 m/s.
  (a) If the whale is stationary in the water, how much time passes before it hears an echo?
  (b) If the whale is moving downward when it emits the sound, will the frequency of the echo be different from the frequency emitted by the whale? If so, will the echo be of higher or lower frequency?

20-11. Mala drops a stone into a well of depth h.
   (a) How much time goes by between her dropping the stone and hearing its splash?
   (b) Calculate the time if the depth of the well is 34 m.

20-12. Allison is traveling on a lake at speed v in her canoe, in the same direction as water waves of speed of $2v$. She counts the waves that pass her as she knifes through the water.
   (a) If her speed is 2.5 m/s and she counts 4 waves in 10 s, what is the wavelength of the waves?
   (b) What is the frequency of the waves as seen by a person on shore?

20-13. Erik visits a canyon. He shouts "hello" and hears his echo soon thereafter.
   (a) How far is Erik from a canyon wall when he hears the echo t seconds after shouting?
   (b) If it takes 6 seconds for him to hear the echo, calculate the distance to the canyon wall.
   (c) If instead Erik is moving toward the wall when he yells, will the echo have the same frequency? Defend your answer.

20-14. Sonar is used to map the ocean floor. Depth is determined by the time it takes for a pulse of sound to reach and then reflect from the ocean floor.
   (a) Find the ocean depth when a pulse is received in time t and the speed of the ultra-sound in water is v.
   (b) Calculate the distance to the ocean floor when a pulse is received 2.0 s after it was emitted. Assume a 1530-m/s speed of ultrasound in water.

20-15. Bats, like dolphins, use ultrasonic sound to locate prey.
   (a) If the time between the ultrasound pulse being sent out and the echo rebounding from a moth and returning is t, how far away is the moth? Assume that the bat is hovering.
   (b) Calculate the distance between the bat and the moth if the time between the original pulse and the echo's return is 0.15 seconds.

20-16. A bat emits a sound pulse when distance d away from an obstacle.
   (a) What is the maximum pulse duration that will allow the bat to hear the echo as distinct from the emitted pulse (assuming that the bat hovers in one place)?
   (b) Why are brief chirps used by bats in navigation?

20-17. Assume that no reflection occurs for objects smaller than a wavelength.
   (a) What is the minimum width detectable by an underwater sonar device that produces waves of speed v and of frequency f?
   (b) Calculate this minimum width for a sonar frequency of 5,000 Hz when sonar waves travel at a speed of 1530 m/s in water.
   (c) Calculate the minimum width for a sonar frequency of 5 MHz.
   (d) To detect smaller objects, should the frequency of the sonar waves be higher or lower? Defend your answer.

20-18. A piano tuner hears 2 beats per second when listening to the combined sound from his tuning fork and a note on the piano being tuned.
   (a) If the fork has a frequency of 440 Hz, what are the possible frequencies of the piano string?
   (b) After a very slight tightening of the string he hears 3 beats per second. Should the string be tightened further or loosened? Defend your answer.

20-19. • Sound waves reach your two ears simultaneously when the
sound source is directly in front of you. But when a sound
source is not in front of you, the direction of the source is
perceived by a slight difference between the arrival times of
the waves at each ear.

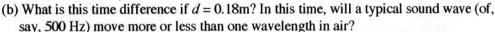

(a) Find the time difference for sound from a source that is at
an angle of 30° relative to the forward direction if the
distance between your ears is d.

(b) What is this time difference if $d = 0.18$m? In this time, will a typical sound wave (of,
say, 500 Hz) move more or less than one wavelength in air?

(c) What happens to the duration of time difference when you rotate your head in a
direction away from the sound source (effectively increasing the angle)?

Show-That Problems

20-20. The wavelength of a 340-Hz tone in air is 1.0 m. Sound travels about 1530 m/s in water. Show that the wavelength of a 340-Hz tone in water is 4.5 m.

20-21. Six seconds after seeing a flash of lightning, the sound is heard. Show that the observer is located about 2 km from the lightning flash.

20-22. Eight seconds after shouting "Hello" in a canyon, the echo returns to you. Show that you are approximately 1400 m from the canyon wall.

20-23. The lowest frequency of sound that an average young person can hear is about 20 Hz. Show that the wavelength of this sound wave is about 17 m (when the speed of sound is 340 m/s).

20-24. The highest frequency of sound that an average young person can hear is about 20,000 Hz. Show that the wavelength of this sound wave is about 1.7 cm.

20-25. After sitting too close to the speakers of too many concerts, the shortest sound waves that Paul can hear are 3.0 cm in wavelength. Show that the highest frequency of sound he can hear is a little over 11,000 Hz.

20-26. Medical ultrasound of frequency 20 MHz is used to diagnose ailments in the body. The speed of ultrasound in body tissue is 1500 m/s. Show that if penetration depth is about 400 wavelengths, the ultrasound can penetrate about 3 cm.

20-27. Some auto-focus cameras use ultrasonic pulses to sense range. For a time delay between emission and detection of 0.010 s, show that the object is 1.7 m in front of the camera.

20-28. A violin and a piano simultaneously sound notes with frequencies 434 Hz and 440 Hz. Show that the beat frequency they hear is 6 Hz.

20-29. A violinist is tuning her instrument to a piano note of 440 Hz detects 4 beats per second. Show that the frequency of the violin can be either 436 Hz or 444 Hz.

20-30. A beat frequency is heard when notes of 240 Hz and 245 Hz are sounded together. Show the same beat frequency occurs when notes of 200 Hz and 205 Hz are sounded together.

Electrostatics

Electrostatics is the study of electric charges at rest. The central rule of electrostatics and electricity in general is: Like charges repel, unlike charges attract. Hence electrons (-) are attracted to protons (+), and are repelled by other electrons. The force of attraction or repulsion between two charges is given by Coulomb's law: $F = k\dfrac{q_1 q_2}{d^2}$, where k is the electrostatic constant, q_1 and q_2 are quantities of charge measured in coulombs, and d is the distance measured in meters between charges.* Note the similarity to Newton's law of gravitation.

Surrounding a charge is an electric field $E = \dfrac{F}{q}$ where F is the force that acts on a test charge q.

By convention, a test charge is a point charge small enough in magnitude to not affect the field being measured. (Note the similarity of the electric field to the gravitational field discussed in Chapter 9, $g = \dfrac{F}{m}$.)

Between the plates of a capacitor the field may be expressed as $E = \dfrac{V}{d}$ where V is the potential difference between the plates and d is the separation distance of the plates. Both the electric force and electrical field are vector quantities.

Sample Problem 1
Two point charges, q_1 and q_2, are separated by a distance d.
(a) Find the magnitude of the force acting between them.

    Solution: From Coulomb's law, $F = k\dfrac{q_1 q_2}{d^2}$.

(b) Calculate the electric force acting between a +2.0 μC charge ($\mu=10^{-6}$) and a –3.0 μC charge separated by 35 cm.

    Solution: Coulombs law: Since k has units of $\dfrac{N \cdot m^2}{C^2}$, charge should be expressed in coulombs and distance in meters.

$$F = k\frac{q_1 q_2}{d^2} = 9.0 \times 10^9 \frac{N \cdot m^2}{C^2} \frac{(2.0 \times 10^{-6}\,C)(-3.0 \times 10^{-6}\,C)}{(0.35\,m)^2} = -0.44\,N. \text{ The negative sign}$$

    indicates that the force is attractive.

---

*Strictly speaking, the equation as written applies to forces between point charges, or between uniformly charged spheres. If the charged objects are small compared with the distance between them, treating the objects as point particles gives an excellent approximation.

Sample Problem 2
Two identically-charged particles are separated by distance r.
(a) At what location between the particles would the force on a positive test charge q_0 be zero?

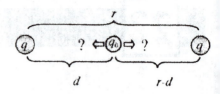

Focus: $d = ?$ where $F_{net} = 0$.

The force on the test charge from the first charge is given by $F = k\dfrac{qq_0}{d^2}$.

The force on the test charge from the second charge is given by $F = k\dfrac{qq_0}{(r-d)^2}$ in the opposite direction. The net force on the test charge will equal zero where it is equally attracted or repelled by the pair of charges. That is, $F_{net} = 0$ when

$k\dfrac{qq_0}{d^2} = k\dfrac{qq_0}{(r-d)^2}$. These two forces will be equal when $d = r - d \implies 2d = r \implies d = \frac{r}{2}$.

This answer makes sense. Since the two charges are identical, the force from each on a third charge will be identical only if it lies the same distance away from each charge. So the test charge experiences zero force at distance $r/2$ from either charge—midway between the two.

(b) Does your answer depend on whether the identical charges are both positive or negative?

Answer: No. At the midpoint between the identical charges, the positive test charge would either be equally attracted to each if they were negative, or equally repelled by each if they were positively charged. In either case $F_{net} = 0$.

Sample Problem 3
A particle of mass m and charge q experiences a force F when located at a point in an electric field.
(a) What is the strength of the electric field at this point?

Answer: $E = \dfrac{F}{q}$, a one-step solution.

(b) How much force would an electron experience at this point?

Solution: From $E = \dfrac{F}{q} \implies F = qE = eE$ (where e is the charge of an electron).

(c) How much acceleration does the electron experience?

Solution: $a = \dfrac{F}{m} = \dfrac{eE}{m}$.

(d) Calculate the acceleration of an electron in an electric field of 540 N/C. The mass of an electron is 9.11×10^{-31} kg, and the charge on an electron is 1.60×10^{-19} C.

Solution: $a = \dfrac{qE}{m} = \dfrac{(1.60 \times 10^{-19}\,\text{C})\left(540\,\frac{\text{N}}{\text{C}}\right)}{9.11 \times 10^{-31}\,\text{kg}} = \mathbf{9.5 \times 10^{13}\,\frac{\text{m}}{\text{s}^2}}$.

Sample Problem 4

Two parallel metal plates are equally and oppositely charged and are separated by distance d. The difference in potential between the plates is V.

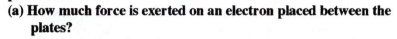

(a) How much force is exerted on an electron placed between the plates?

Focus: $F = ?$

From $E = \dfrac{F}{q} \Rightarrow F = qE = eE$,

where e is the charge of an electron. The pair of plates comprise a capacitor, with electric field between the plates $E = \dfrac{V}{d}$. So $F = e\left(\dfrac{V}{d}\right) = \dfrac{eV}{d}$.

This force is the same anywhere between the plates since the electric field is constant between the plates of a capacitor.

(b) How much work would be required to move an electron from the positively charged plate to the negatively charged plate?

Focus: $W = Fd = \left(\dfrac{eV}{d}\right) d = eV.$

Note the units check. Work is in joules, e is in coulombs, and V has units joules/coulomb, so eV has units joules. Checking units assures that you're either on, or not on, a correct track!

Problems

22-1. A couple of pennies are each given an electrical charge. One is given a charge $-q$ and the other a charge $+2q$. When the pennies are brought together and touch, the charges redistribute and equalize over their surfaces.
(a) What is the final charge on each penny?
(b) Calculate the final charge on each penny if $-q$ is -30 μC (-30 × 10^{-6} C).

22-2. Two spherical inflated rubber balloons each have the same amount of charge spread uniformly on their surfaces. They repel each other with a force F when their centers are distance d apart.
(a) What is the charge on each balloon?
(b) If the repelling force is 2.5 N and the distance between the balloon centers is 0.30 m, calculate the charge on each balloon.

22-3. In accord with Coulomb's law, a force of attraction or repulsion occurs between pairs of charged particles.
(a) Calculate the force between a particle with charge -5.0 μC and a particle with charge $+3.0$ μC separated by a distance of 0.50 m.
(b) Calculate the force between a particle with charge $+5.0$ μC and a particle with charge -3.0 μC separated by a distance of 0.50 m.
(c) Does the magnitude of calculated forces depend on the signs of the charges? Explain.

22-4. Consider two small charged objects, one with charge q and the other of unknown charge. When they are separated by a distance d, each exerts a force F on the other.
(a) What is the charge of the second object?
(b) Calculate the charge of the second object if the attractive force between the two charges is 2.8 N, the charge on the first object is 15 μC, and the distance between them is 1.2 m.

22-5. Consider the pair of protons in the nucleus of a helium atom, each with charge e (same as the electron, but positive) and mass m_p. Assume their center-to-center distance to be d.
(a) Write an expression for what their initial acceleration would be if they were free to accelerate away from each other. (We'll see in Chapter 33 that an attractive force, the strong nuclear force, prevents protons from accelerating away from each other.)
(b) Calculate the hypothetical initial acceleration of each proton in the pair without the presence of the strong nuclear force. The mass of a proton is 1.67×10^{-27} kg, and assume that the protons in a helium nucleus are about 3×10^{-15} m apart.

22-6. A particle of charge q is at a distance d from the center of the charged dome of a Van de Graaff generator. Another particle of charge $9q$ experiences the same force at a different distance from the center of the dome.
(a) Find the distance of the particle of charge $9q$ from the dome in terms of distance d.
(b) Does your answer depend on the sign of the charges? Defend your answer.

22-7. A particle of mass m with charge q is held in the horizontal electric field to one side of the charged spherical dome of a Van de Graaff generator. When released, the particle's horizontal component of acceleration is equal to eight times the acceleration due to gravity.
(a) What is the electric field at this location?
(b) What would be the particle's initial horizontal component of acceleration if it were instead released at a distance twice as far horizontally from the center of the charged sphere?

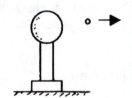

22-8. A particle of charge q is held near the charged dome of a Van de Graaff generator. When the particle is released it undergoes acceleration a away from the generator. Ignore gravity.
(a) At what distance A from the center of the dome would the particle's acceleration be one-quarter as much as its initial acceleration a? (Answer in terms of the original distance from the generator's center.)
(b) At what distance B from the center of the generator would acceleration of the particle be one-tenth a?
(c) Is the particle's acceleration when passing point A greater, the same, or less than its acceleration when passing point B?
(d) Is the particle's *speed* when passing point A greater, the same, or less than its speed when passing point B?
(e) Why are your answers to (c) and (d) different?

22-9. A hydrogen atom is composed of a proton and an electron with an average separation distance of 5.3×10^{-11} m. The magnitude of charge on each is 1.6×10^{-19} C.
   (a) Calculate the magnitude of force between the proton and the orbiting electron at this average distance.
   (b) The proton pulls the electron into an orbit with the force you calculated above. How much force, in comparison, does the electron exert on the proton?

22-10. Assume the orbiting electron of the previous problem traces a circular orbit. Then the electric force would be a centripetal force.
   (a) What is the centripetal acceleration of the electron?
   (b) What is its speed?

22-11. After scuffing your shoes on a hotel rug and gaining a charge $-q$, you see a charming someone with an equal and opposite charge $+q$ across the hall a distance x away.
   (a) Approximately what is the strength of your electrical attraction to the charming person?
   (b) What is the charming person's attraction to you?
   (c) Calculate the strength of the attraction if you each have identical magnitudes of charge $50\mu C$ while you're 1.5 m apart.

22-12. Two identical particles with charge q are separated by a distance d and repel each other with a force F.
   (a) By how much does the force change if *each* charge is doubled, while the distance between them remains unchanged?
   (b) By how much does the force change if *each* charge is doubled, while the distance between them also doubles?
   (c) By how much does the force change if the charge of only *one* particle doubles, and the distance between them doubles?

22-13. A 1.0-μC particle charge feels a downward electrical force of 1.0×10^{-4} N at Earth's surface.
   (a) What is the strength of the electric field at Earth's surface?
   (b) What is the charge of Earth?
   (c) How many excess electrons does this represent?

22-14. An electric field occupies the space around every electric charge.
   (a) What is the electric field 0.010 m from an isolated proton?
   (b) What is the electric field 0.010 m from an isolated helium nucleus (an alpha particle, with twice the charge of a single proton)?
   (c) What is the electric field 0.010 m from an isolated neutral helium atom?

22-15. The electric field at distance d from a point charge q is 9.80 N/C. The field strength decreases with distance via the inverse-square law.
   (a) At what distance from the point charge is the electric field 0.098 N/C? (Express your answer as a multiple of the initial distance.)
   (b) At what distance from the point charge is the field 980 N/C?

22-16. A certain parallel-plate capacitor has a plate separation d and a potential difference between the plates of V. The electric field between the plates is uniform.
(a) What is the strength of the electric field E?
(b) Calculate the strength of the electric field if the distance apart is 0.05 m and a voltmeter shows a 100-V potential difference between the plates. (Note that the unit V/m is equivalent to N/C.)
(c) If the acceleration of a charged particle between the plates is a when it is halfway between the plates, why will the particle experience the same acceleration if it is only one-quarter the distance between the plates? Defend your answer.

22-17. An electron is placed in the region between the parallel plates of a capacitor.
(a) What is the magnitude of force on the electron if the uniform field is 1000 N/C?
(b) Which will be constant for the electron in this field, acceleration or speed? Defend your answer.

22-18. A simple electron gun consists of two oppositely-charged plates, the negative one called the cathode and the positive one called the anode. Electrons are "boiled" off the cathode and accelerate toward the anode. A hole in the anode allows an electron beam to emerge from the device.

(a) If the potential difference between anode and cathode is 100 V, calculate the kinetic energy of an electron emerging from the gun.
(b) Calculate the velocity of the electrons in the beam that emerges through the hole.

22-19. Ink-jet printers spray charged drops of ink onto paper. An electric field in the printer head produces a force on the ink drops.
(a) Find the electric field strength that produces a force of 2.8×10^{-4} N on a drop having a charge of 1.6×10^{-10} C.
(b) If the electric field were increased in strength by 10%, by what factor would the force on the charged drops increase?

22-20. Particles with charges q and $4q$ have the same mass and are moving in the same uniform electric field E.
(a) How do the forces exerted by the field on these two particles compare?
(b) How do the accelerations of these particles compare?
(c) How would the accelerations compare if the particle with charge $4q$ had four times the mass of the other particle?

22-21. In 1909 Robert Millikan was the first to find the charge of an electron in his now-famous oil drop experiment. He levitated charged drops of oil in an electric field so that an upward electrical force balanced the gravitational force on a drop.

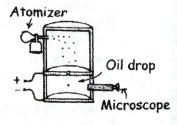

(a) If a drop of mass 1.1×10^{-14} kg remains stationary in an electric field of 1.68×10^{5} N/C, what is the charge on the drop?
(b) How many extra electrons are on this drop?

22-22. An electric field does work to move charges placed in the field.
- (a) How much work does an electric field E do when a charge q is moved a distance x in the direction of the field?
- (b) Calculate the work done by a 7500 N/C electric field when a 55 μC charge moves 42 cm in the direction of the field.

22-23. The potential difference between a particular storm cloud and the ground is 100 million volts.
- (a) If a charge of 2.5 C flashes in a bolt from the cloud to Earth, how big is the change of potential energy of the charge?
- (b) How big is the change in potential energy for a 5-C flash between the same cloud and ground?

22-24. When a small sphere of mass m and charge q_1 is suspended from a vertical spring the spring stretches somewhat due to the gravitational force acting on m (recall Hooke's law: $F = kx$, where the stretch of a spring is proportional to the force on it). The spring stretches an additional distance x when another sphere of charge q_2 is brought beneath it. The final distance between the centers of the spheres is d.
- (a) With how much force are the spheres attracted? (Hint: Your answer can be given without using any property of the spring.)
- (b) What is the spring constant k of the spring (don't confuse spring constant k with the electrostatic constant k in Coulomb's law)?

22-25. • Three identically charged particles are placed on the tips of an equilateral triangle.
- (a) Draw the forces acting on each particle due to the other two, indicating the directions and relative magnitudes.
- (b) Determine the relative magnitude and angle of the resultant force on any particle.

Show-That Electrostatics Problems

22-26. The charge on an electron is 1.60×10^{-19} C.
Show that 6.25×10^{18} electrons have a collective charge of 1 C.

22-27. In walking across a carpet you might build up a net negative charge of 50 μC.
Show that you have acquired about 310 million million excess electrons on your body.

22-28. The distance between two charged particles is doubled.
Show that the electrical force between them is reduced to one-fourth of its previous value.

22-29. The distance between two charged particles is halved.
Show that the electrical force between them is four times its previous value.

22-30. The distance between two charged particles is doubled, and the charge on one of the particles is halved.
Show that the electrical force between them is one-eighth of its previous value.

22-31. A test charge at a distance d from a charge q experiences a force F.
Show that the electric field in the location of the test charge is $E = kq/d^2$.

22-32. A particle of mass m and charge $2q$ is held in a uniform electric field E.
Show that the work needed to move the charge a distance x against the electric field is $2qEx$.

22-33. An isolated proton is acted on by an electrical force of 4.8×10^{-14} N.
Show that the magnitude of the electric field at the proton's location is 3.0×10^5 N/C.

22-34. A charge of mass m and charge q accelerates from rest from one plate of a parallel plate capacitor. The distance between the plates is d and the potential difference between them is V.
Show that the acceleration of the charge is given by $a = \dfrac{qV}{md}$.

22-35. A proton has weight and is pulled downward by gravity. An electric field could push it upward.
Show that the magnitude of a vertical electric field that would just support the weight of a proton near Earth's surface is 1.0×10^{-7} N/C upward.

22-36. Millikan discovered that a tiny charged drop would remain suspended in an electric field that counteracted the gravitational field.
Show that for a drop of mass m carrying charge q held motionless in a vertical electric field E, the following equation holds true: $qE/mg = 1$.

22-37. As the charge on a drop of given mass m gets smaller, the electric field required to hold the drop motionless gets larger.
Show that the maximum needed field is $E = mg/e$ when the drop carries a single quantum unit of charge, e.

22-38. A square with sides L has identical charges q on three of its corners.
Show that the magnitude of the electric field at the center of the square is $2kq/L^2$.

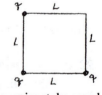

22-39. A square with sides L has identical charges on three of its corners.
Show that the electric field at the remaining corner has a magnitude approximately equal to $1.9kq/L^2$.

22-40. The electron gun in the picture tube of yesterday's television sets accelerate electrons across a potential difference of 2000 V.
Show that the final speed of such an accelerated electron is about 2.7×10^7 m/s.

23 Electric Current

Much of the confusion in understanding electricity is a failure to correctly distinguish among the concepts of *current, voltage, energy,* and *power.* The problems herein may help clarify these distinctions.

Equations for this chapter

$I = q/t$ [Current = electric charge/time, or charge that flows per unit time]
$V = Energy/q$ [Voltage (electric potential) = energy/charge, or energy per unit charge]
$I = V/R$ [Ohm's law; current = voltage/resistance]
$P = IV$ [Power = current × voltage]

Units of measurement

C, coulomb: quantity of charge = charge on 6.25×10^{18} electrons
V, volt = joules/coulomb: voltage, or electric potential
A, ampere = coulombs/second: current
Ω, ohm = volts/amp: resistance R
W, watt = joules/second: power

> Current is a flow of charge (I), pressured into motion by voltage (V), and hampered by resistance (R).

Sample Problem 1
A current I flows through an automobile headlight.
(a) How many coulombs of charge flow through its filament in time t?

Focus: $q = ?$

From the definition of current, $I = \dfrac{\text{charge}}{\text{time}} = \dfrac{q}{t} \Rightarrow q = It.$

In SI units q is in coulombs, I is in amperes, and t is in seconds.

(b) Calculate the amount of charge in coulombs that flow through the lamp filament in 30 minutes when the current is 4.0 A.

Solution: $q = It$. Recognizing that 1 A = 1 C/s, $q = \left(4.0\frac{C}{s}\right)\left(30 \text{ min} \times \frac{60\,s}{1\,min}\right) = \textbf{7200 C}.$

(c) Current has direction in a circuit, conventionally from the positive to the negative terminal of the connected battery. Can we therefore say that current is a vector quantity?

Answer: No. The current in a wire is always along the length of the wire, whether the wire is straight or curved. No single vector could describe motion along a curved path. We arbitrarily call the positive direction that in which positive charge would flow (even though the actual flow of negative electrons is in the other direction).

Sample Problem 2

An automobile battery, rated at 12 volts, provides the electrical pressure in an electric circuit.

(a) Find the amount of energy per coulomb of charge making up the current in the circuit?

Focus: $\dfrac{\text{Energy}}{\text{coulomb}} = ?$

Voltage is defined as energy per charge. So by definition, $12 \text{ volts} = \dfrac{12 \text{ joules of energy}}{\text{coulomb of charge}}$.

(b) How much energy is provided to the circuit when q coulombs of charge pass through the battery?

Solution: From $V = \dfrac{\text{Energy}}{q} \Rightarrow \text{Energy} = qV$.

(c) Calculate the energy provided to the circuit when 8 coulombs of charge pass through a 12.0-V circuit.

Solution: $\text{Energy} = qV = (8.0 \text{ C})\left(12.0\tfrac{\text{J}}{\text{C}}\right) = \textbf{96 J}$.

Sample Problem 3

In the simple circuit shown to the right we see a battery and a single lamp. The battery voltage is V and the resistance of the lamp is R. The internal resistance in the battery is small enough to ignore.

(a) How much current is in the lamp when the switch is closed?

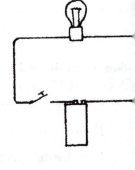

Focus: $I = ?$
The current in the lamp depends on the magnitude of the electric pressure, V, and the magnitude of the resistance opposing the flow of charge in the circuit.

The solution is given by Ohm's law, $I = \dfrac{V}{R}$.

(b) Calculate the current in the lamp if it has a resistance of 1 Ω and the battery voltage is 6 V.

Solution: $I = \dfrac{V}{R} = \dfrac{6 \text{ V}}{1 \text{ }\Omega} = \textbf{6 A}$.

(c) Why does the filament of a lamp glow, but not the connecting wires?

Answer: Energy is delivered and dissipated in locations of circuit *resistance*. In a single-lamp circuit almost all of the resistance is in the filament of the lamp, where energy is converted to heat and light. Resistance generally increases with increases in temperature, so resistance is greater once the lamp filament heats up. The resistance offered by the connecting wires and the battery interior are usually negligible compared with the lamp's resistance. That's why, for simplicity, we ignore all resistance in the circuit except that of the lamp.

Sample Problem 4
This is the same circuit, but with an additional identical lamp.
(a) How much current is in the lamps when the switch is closed?

Focus: $I = ?$ These bulbs are connected in series. Since each bulb has a resistance R, the total resistance in the series circuit is $2R$.

The solution is again given by Ohm's law, $I = \dfrac{V}{2R}$.

The current in this circuit is half what it was in the single-lamp circuit. (We're making the assumption that the resistance is independent of temperature effects.)

(b) Calculate the current in the lamps if each has a resistance of 1 Ω and the battery voltage is 6 volts.

Solution: $I = \dfrac{V}{2R} = \dfrac{6\text{ V}}{2\ \Omega} = 3\text{A}.$

That's 3 amps through each lamp, and through the battery. The 6 volts across the circuit divides between the two lamps.

(c) Can we say that since the current in each lamp is 3 A that the total current in the circuit is 6 A?

Answer: No! Every time an electron enters the lamp filament another one must leave. They don't "pile up" anywhere in the circuit. So if 3 coulombs of charge per second enter and leave the first bulb, 3 coulombs per second enter and leave the second bulb as well.

(d) Through which of the two lamps does current first flow?

Answer: Current is established in both lamps at once. It doesn't matter which lamp is closer to which terminal of the battery. When the switch is closed, an electric field is established in all parts of the circuit at practically the same time. All free electrons in every part of the circuit, including the battery interior, move in unison similar to the command "forward march" that makes each member of a marching band begin stepping at the same time.

Sample Problem 5
Here we have the same two lamps of the previous problem, but connected differently. We say they are connected in parallel.
(a) Calculate the current in each lamp when the switch is closed.

Focus: $I = ?$
In this arrangement voltage does not divide between the bulbs as in a series circuit. Careful inspection will show that each lamp is connected across the terminals of the 6-V battery, so each lamp is "energized" with a full 6 volts. Thus the current through each lamp is

$I = \dfrac{V}{R} = \dfrac{6\text{ V}}{1\ \Omega} = \textbf{6 A}.$

The greater the current in a lamp, the brighter it glows.

(b) What is the current in the battery?

Answer: Current in each lamp is 6 A (6 coulombs per second), which means that current in the battery must be the sum of these, 12 amps (12 coulombs per second).

(c) Doesn't an increased current in the battery defy Ohm's law?

Answer: Not at all. In accord with Ohm's law, the battery supplies twice the current to the circuit because there are twice as many paths available for current to flow. The *equivalent resistance* of the circuit is half that of a single-lamp circuit.* Similar to increased checkout lines in a supermarket, more branches in a parallel circuit reduce resistance and allow for a greater flow.

Sample Problem 6

Power is energy dissipated per time. Each of the circuits discussed dissipates energy.
(a) Calculate the power dissipated in the single lamp circuit.

Solution: $\text{Power}\left(\text{watts}, \frac{J}{s}\right) = \text{Current}\left(\text{amps}, \frac{C}{s}\right) \times \text{Voltage}\left(\text{volts}, \frac{J}{C}\right)$. The power in the circuit comes from the energy provided by the battery and supplied to the charges passing through it. For the single bulb

$$P = IV = \left(6\frac{C}{s}\right)\left(6\frac{J}{C}\right) = 36\frac{J}{s} = \textbf{36 W}.$$

(b) Calculate the power dissipated in the two-lamp series circuit.

Solution: 3 amps (3 coulombs per second) pass through the battery. The battery provides 6 joules of energy per coulomb (that is, 6 volts.) So $P = IV = \left(3\frac{C}{s}\right)\left(6\frac{J}{C}\right) = 18\frac{J}{s} = \textbf{18 W}$.

If you prefer to focus on the individual bulbs: A coulomb of charge leaves the battery with 6 joules of potential energy. The energy is distributed equally between the two identical bulbs, 3 joules per coulomb transformed to heat and light in each one. Since 3 amps (3 coulombs per second) pass through both lamps, the power dissipated in each lamp is

$$P = IV = \left(3\frac{C}{s}\right)\left(3\frac{J}{C}\right) = 9\frac{J}{s} = 9 \text{ W, for a total of } \textbf{18 W} \text{ all together.}$$

(c) Calculate the power dissipated in the two-lamp parallel circuit.

Solution: 12 amps (12 coulombs per second) pass through the battery. The battery provides 6 joules of energy per coulomb (that is, 6 volts.) So $P = IV = \left(12\frac{C}{s}\right)\left(6\frac{J}{C}\right) = 72\frac{J}{s} = \textbf{72 W}$.

If you prefer to focus on the individual bulbs: Each bulb in the parallel circuit is just like the individual bulb in the single bulb circuit. A coulomb of charge leaves the battery with 6 joules of potential energy and 6 amps (6 coulombs per second) pass through each lamp. The power dissipated in each lamp is $P = IV = \left(6\frac{C}{s}\right)\left(6\frac{J}{C}\right) = 36\frac{J}{s} = 36 \text{ W, for a total of } \textbf{72 W}$ for the two together.

---

*From the battery's perspective, it is a 6-V battery putting out 12 A. From Ohm's law, $R = V/I = 6$ V/12 A = 0.5 Ω. As far as the battery is concerned, the two 1-Ω resistors in parallel are *equivalent* to and indistinguishable from a single 0.5-Ω resistor.

(d) Which circuit will drain the battery the quickest? The slowest?

Answer: The two bulbs in parallel draw the most power and will give off the most light. They will draw a large current, but for a relatively shorter period of time. The two bulbs in series provide the greatest resistance of the three circuits, so the current through them will be the smallest and the battery will last the longest. Indeed, most batteries are rated in *amp-hours*, the product of the current they provide and the number of hours for which they can provide it. The two lamps in series will stay lit four times as long as the two lamps in parallel because the series circuit draws only one-fourth as much current.

Sample problem 7
A 2-Ω and a 6-Ω resistor are connected in parallel to a 9-volt battery.
(a) How much current runs through the battery?

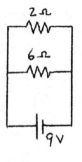

Focus: $I_{battery} = ?$
Each resistor "sees" an electric pressure of 9 volts across it.

The current through the 2-Ω resistor is $I = \dfrac{V}{R} = \dfrac{9\,V}{2\,\Omega} = 4.5\,A.$

The current through the 6-Ω resistor is $I = \dfrac{V}{R} = \dfrac{9\,V}{6\,\Omega} = 1.5\,A.$

The total current supplied by the battery is the sum of these two currents:
$I_{battery} = 4.5\,A + 1.5\,A = \mathbf{6.0\ A}.$

(b) What single resistor connected to the battery would allow the same current through the circuit as the parallel connection of the 2-Ω resistor and 6-Ω resistor?

Focus: R = ?
We want to find the resistance that will give 6.0 A in a 9.0-V circuit.

From $I = \dfrac{V}{R} \;\Rightarrow\; R = \dfrac{V}{I} = \dfrac{9.0\,V}{6.0\,A} = \mathbf{1.5\,\Omega}.$ So we can say, resistance-wise, that a 2-Ω resistor

and 6-Ω resistor in parallel are equivalent to a single 1.5-Ω resistor. In general the *equivalent resistance* of two resistors in parallel is their product divided by their sum. That is,

$$R_{equivalent,\,parallel} = \frac{R_1 R_2}{R_1 + R_2}.^{*}$$

In this situation, $R_{eq} = \dfrac{(2\,\Omega)(6\,\Omega)}{2\,\Omega + 6\,\Omega} = \mathbf{1.5\,\Omega}.$

(c) What would be the equivalent resistance of the 2-Ω and 6-Ω resistor connected in series?

The resistances in a series circuit just add up. So $R_{eq} = R_1 + R_2 = 2\,\Omega + 6\,\Omega = \mathbf{8\,\Omega}.$

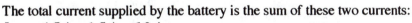

* More generally, the equivalent resistance for n resistors in parallel is given by $\dfrac{1}{R_{eq}} = \dfrac{1}{R_1} + \dfrac{1}{R_2} + \cdots + \dfrac{1}{R_n}.$

Sample Problem 8
Three 2-Ω resistors and one 3-Ω resistor are connected in the circuit shown.
(a) What is the equivalent resistance of the circuit?

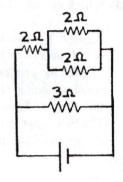

Focus: $R_{eq} = ?$
We can break the circuit into steps as shown below. In Step 1 note the pair of 2-Ω resistors in parallel have an equivalent resistance of 1 Ω

(using $R_{equivalent, parallel} = \dfrac{R_1 R_2}{R_1 + R_2}$). Then in Step 2 note that the 2 Ω

and equivalent 1 Ω are in series and equal 3 Ω. So the circuit becomes a pair of 3-Ω resistors in parallel. In Step 3 we see the equivalent resistance of a pair of 3-Ω resistors in parallel is 1.5 Ω. So the equivalent resistance of the circuit is **1.5 Ω**.

Step 1   Step 2   Step 3

(b) If the battery is 12 V, calculate the current in the battery?

Answer: The current in the battery is the line current in the circuit,

which by Ohm's law is $\dfrac{V}{R_{eq}} = \dfrac{12\ \text{V}}{1.5\ \Omega} = \textbf{8 A}$.

> Current in the line connected to the battery is the same as the current in the battery!

Problems

23-1. A wire carries a current I. An amount of charge q drifts past a point in the wire during some time interval.
(a) How much time does it take for q to drift past a point in the wire?
(b) Calculate the time it takes for 25 C to drift through a wire carrying 12.0 A.

23-2. An automobile engine's starter motor draws a current I.
(a) How much charge q flows through the motor if the motor runs for time t?
(b) Calculate the quantity of charge that flows in a starter motor when it draws 45.0 A for 1.25 s.
(c) How many electrons does this represent?

23-3. A silver wire transfers a charge q in time t.
   (a) What is the current in the wire?
   (b) Calculate the current in the wire when 420 C is transferred in 60 min.

23-4. When a clothes iron is connected to a voltage source, its heating element draws a current.
   (a) Calculate the amount of current the iron draws when connected to 120 V if the resistance of the heating element is 18 Ω.
   (b) If the voltage were doubled, what would be the current?

23-5. Some materials are better electrical conductors than others. The *resistivity* of a material tells us how well or how poorly a material conducts electricity. A material with a higher resistivity is a poorer conductor. The resistance of a wire is directly proportional to the resistivity of the material from which the wire is made and is also directly proportional to the length of the wire. Increasing either the resistivity of the material or the length of the wire increases the wire's resistance. That is, $R \sim resistivity \times length$. (The resistance is also inversely proportional to the cross-sectional area of the wire, but we won't use that dependence in this problem.) Consider two lengths of wire of the same thickness and the same resistance, one made of aluminum and the other made of copper. Aluminum has 1.6 times greater resistivity than copper for the same temperature.
   (a) How much longer is one wire compared with the other?
   (b) If 20.0 m of aluminum wire has the same resistance and same thickness as the copper wire, how long is the copper wire?

23-6. Two lengths of wire have the same thickness. One is tungsten and the other is silver. Tungsten has 3.8 times greater resistivity than silver at the same temperature. (See information at the beginning of Problem 23-5.)
   (a) If two lengths of silver and tungsten wire of the same thickness have the same resistance, how much longer is one compared with the other?
   (b) If 20.0 m of tungsten wire has the same resistance and the same thickness as the silver wire, how long is the silver wire?

23-7. Two lengths of wire have the same thickness. One is nichrome and the other is silver. Nichrome has 68 times greater resistivity than silver at the same temperature. (See information at the beginning of Problem 23-5.)
   (a) If two lengths of nichrome and silver wire of the same thickness have the same resistance, how much longer is one compared with the other?
   (b) If 20.0 m of nichrome wire has the same resistance and the same thickness as the silver wire, how long is the silver wire?

23-8. Four identical resistors, each of resistance R, are combined.
   (a) What is their equivalent resistance when all are connected in series?
   (b) What is their equivalent resistance when all are connected in parallel?
   (c) What is the equivalent resistance of two parallel branches, each with two resistors in series?

23-9. Circuit A with a 12-V battery has 3 identical 2-Ω resistors. The switch is closed.
(a) What is the voltage across each resistor?
(b) What is the current in each resistor?
(c) What power does the battery deliver to the circuit?

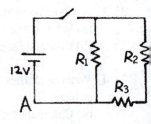

23-10. Circuit B with a 12-V battery has 3 identical 2-Ω resistors. The switch is closed.
(a) What is the voltage across each resistor?
(b) What is the current in each resistor?
(c) What power does the battery deliver to the circuit?
(d) What is the voltage across each resistor when the switch is open?
(e) What is the current in each resistor when the switch is open?

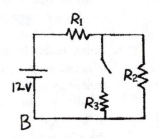

23-11. A 16-Ω loudspeaker and an 8-Ω loudspeaker are connected in series across the terminals of an amplifier.
(a) What is the equivalent resistance of the two speakers?
(b) What is their equivalent resistance if they are connected, more properly, in parallel?

23-12. Circuit C with the 12-V battery has 4 identical 2-Ω resistors. The switch is closed.
(a) What is the voltage across each resistor?
(b) What is the current in each resistor?
(c) What power does the battery deliver to the circuit?
(d) What is the voltage across each resistor when the switch is open?
(e) What is the current in each resistor when the switch is open?

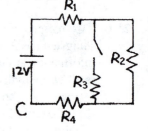

23-13. Circuit D is powered with a 12-V battery, but has resistances of different values as indicated.
(a) What is the voltage across each resistor?
(b) What is the current in each resistor?
(c) What power does the battery deliver to the circuit?

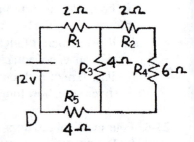

23-14. Power is required to run electric motors.
(a) Calculate the power input to a motor connected to a 120-V circuit when it draws 8.0 A.
(b) Calculate the energy used when the motor operates for 8.0 hours.

23-15. The current drawn by a light bulb depends on its power rating.
(a) How much current is drawn by a 60-W bulb rated for 120 V?
(b) What is the resistance of the filament in the bulb?
(c) How much energy lights a 60-W bulb left on for a time of 8.0 hours?

23-16. A strip of a certain wire is used as a fuse in a circuit whose current is to be limited to I amperes. The fuse melts after it has received E joules of energy in t seconds.
(a) What power is going to the fuse at the time of melting?
(b) What is the resistance of the fuse wire?
(c) If the current in the circuit is to be limited to 10 A, and the fuse melts when it receives 9.0 J in 1.0 s, calculate the resistance of the fuse.

23-17. A hair dryer requires power for operation.
(a) Calculate the current drawn by a 1500-W hair dryer connected to a 120-V circuit.
(b) What current will it draw if incorrectly connected to a 240-V circuit, and why might it burn out?
(c) Which stays the same, the resistance of the dryer or the power it dissipates?

23-18. Six identical lamps are connected in series across a 120-V line.
(a) What is the voltage across each one?
(b) If the current is 0.50 A, calculate the resistance of each lamp.
(c) Calculate the power dissipated by each lamp.

23-19. Six identical lamps are connected in parallel in a 120-V circuit.
(a) What is the voltage across each lamp?
(b) If the total current in the circuit is 5.0 A, calculate the resistance of each lamp.
(c) Calculate the power dissipated by each lamp.

23-20. • A pair of resistors connected in series to a 120-V line uses one-fourth the power they use when connected in parallel to the same line.
(a) Why does it make sense the resistors use more power when connected in parallel than when connected in series?
(b) Calculate the resistance of one of the resistors if the other is 1800 Ω.

23-21. You have a 40-W bulb and a 100-W bulb, both rated for a 120-V circuit.
(a) What is the resistance of each bulb?
(b) What is the equivalent resistance when they are connected in series?
(c) Calculate the power dissipation in each bulb when connected in series.
(d) What is the equivalent resistance when they are connected in parallel?
(e) In which type of circuit will the 40-W bulb be brighter than the 100-W bulb?

23-22. The circuit shown has three identical lamps, each of resistance R, in parallel across a battery of voltage V.
(a) If the battery voltage is 6 V, what is the voltage across each of the three branches?
(b) If each lamp has a resistance of 2 Ω, calculate the current in each branch.
(c) Find the equivalent resistance of the circuit.
(d) Calculate the current in the battery.

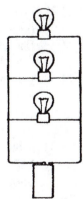

Show-That Problems

23-23. A wire carries a current of 10 A.
Show that the time it takes for a charge of 60 C to pass by a point in the wire is 6 seconds.

23-24. A current of 0.03 A through the human body is usually fatal.
Show that when in contact with 120 V, a minimum body resistance of 4000 Ω is crucial.

23-25. When the human body is dry it has a resistance of about 120,000 Ω. Suppose a human body is in contact with 120 V.
Show that a current of 0.001 A in the body will be produced.

23-26. A 1.5-V cell with an internal resistance of 0.20 Ω is connected to a flashlight bulb having a resistance of 1.8 Ω.
Show that the current in the bulb is 0.75 A.

23-27. A 15-A fuse melts after it has received 12.0 J of energy in 1 s.
Show that the resistance of the fuse is 0.053 Ω.

23-28. A 12-V battery energizes two bulbs in series, one with resistance 12 Ω and the other with resistance 18 Ω.
Show that the power delivered by the battery is 4.8 W.

23-29. A 12-V battery energizes two bulbs in parallel, one with resistance 12 Ω and the other with resistance 18 Ω.
Show that the power delivered by the battery is 20 W.

23-30. Three resistors of 6 Ω, 8 Ω, and 16 Ω are connected in series to a 12-volt battery of negligible internal resistance.
Show that the current in the circuit is 0.4 A.

23-31. Three resistors of 6 Ω, 8 Ω, and 16 Ω are connected in series to a 12-volt battery of negligible internal resistance.
Show that the power expended in the circuit is 4.8 W.

23-32. Five identical lamps are connected in series to a source that provides 550 volts, which produces a current of 1.1 A in the lamps.
Show that the total resistance of all five lamps is 500 Ω.

23-33. The same five identical lamps of the previous problem are connected in parallel with a source that provides 550 volts.
Show that the current in the source is 28 A.

23-34. A 60-Ω heater and a 144-Ω lamp are connected in parallel to a generator that supplies 120 V.
Show that the current in the generator is 2.8 A.

23-35. A dozen 120-Ω lamps are connected in parallel to a 120-V house line.
Show that the line current will not blow a 15-A fuse.

23-36. A dozen 120-Ω lamps are connected in parallel to a 120-V house line.
Show that the power that illuminates each lamp is 120 W.

23-37. Four 600-W heaters are connected in parallel to a 120-V line. Each heater can be turned on or off independently.
Show that the increase in line current as each heater is turned on is 5 A.

23-38. The total current in a parallel circuit of n branches is given by the sum of the currents in its branches. That is, $I_{total} = I_1 + I_2 + I_3 \ldots I_n$. Since the voltage V is the same across all branches, Ohm's law tells us $\dfrac{V}{R_{equivalent}} = \dfrac{V}{R_1} + \dfrac{V}{R_2} + \dfrac{V}{R_3} + \ldots \dfrac{V}{R_n}$.
Show that dividing both sides of the equation by V gives an expression for the equivalent resistance of resistors in parallel.

23-39. The equivalent resistance for resistors of n branches in parallel can be found from *the reciprocal rule*: $\dfrac{1}{R_{equivalent}} = \dfrac{1}{R_1} + \dfrac{1}{R_2} + \dfrac{1}{R_3} + \ldots \dfrac{1}{R_n}$.
Show that for two resistors in parallel that $R_{equivalent} = \dfrac{R_1 R_2}{R_1 + R_2}$. This is often called the *product-over-sum rule*, which applies only to *pairs* of resistors in parallel.

23-40. You have three 60-Ω resistors.
Show how to combine them to produce an equivalent resistance of 20 Ω.

23-41. You have three 60-Ω resistors.
Show how to combine them to produce an equivalent resistance of 90 Ω.

23-42. A 60-Ω resistor and a 20-Ω resistor are connected in parallel. This combination is then connected in series with a 33-Ω resistor. The circuit is connected to a 24-V battery.
Show that the current in the 33-Ω resistor is 0.5 A.

23-43. A 100-W lamp is left on for 24 hours.
Show that the energy expended in keeping it lit is 8.64 MJ.

23-44. A 100-W stereo system is connected to a 120-V outlet.
Show that the current drawn by the system is 0.83 A.

23-45. A string of 50 party lights connected in series operates at 120 V.
Show that the resistance of each bulb is 24 Ω if 0.10 A flows through the string of bulbs.

We have seen that an electric charge is surrounded by an electric field that can affect other charges. Likewise, a mass has a gravitational field that affects masses around it. In this chapter we look at the effects of *magnetic fields* that surround magnets and moving charges, which can affect other magnets and moving charges. Just as the direction of the electric field points in the direction of the force acting on a positively-charged particle, the magnetic field points in the direction of the force that would act upon a (theoretical) north magnetic pole.

When magnetic fields interact with moving charged particles, the force experienced by the particles depends upon three factors:

    (1) The charge q (its sign and magnitude).
    (2) The velocity v of the charged particle.
    (3) The magnetic field vector B (its strength and direction).

Thus, our magnetic force equation looks something like

    Magnetic force $\approx$ charge $\times$ velocity $\times$ magnetic field strength.

The magnetic force is peculiar in that it acts in a direction perpendicular to both the velocity vector and magnetic field vector. The magnitude of the force depends on the component $v_\perp$ of the velocity vector that is perpendicular to the magnetic field vector B. In this book we will only consider particles moving perpendicular to the magnetic field.

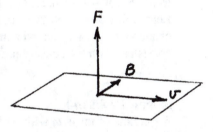

The magnetic force equation is $F = q v_\perp B$.

The SI unit for B is the tesla (T), which is 1 newton per ampere-meter. So 1 T = 1 N/(A·m), or equivalently, 1 N/(C·m/s).

Notice in the above diagram that the force vector F is perpendicular to both $\vec{v}$ and $\vec{B}$. The conventionally correct direction for F can be determined using the *right-hand rule*. The diagram at right shows the relative directions of v, B and F for a positive charge (forefinger pointing in the direction of v, middle finger in the direction of B, and thumb in the direction of the force F).[*]

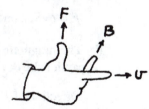

When a current flows through a wire, the moving charges in it produce a magnetic field. Another version of the right-hand rule shows the direction of the magnetic field B around a current-carrying wire.[†] Here you grab the (insulated!) wire with your right hand, with your thumb pointing in the direction of the (positive charge carrier) current. The direction of your fingers curling around the wire indicates the direction of the magnetic field circling around the wire.

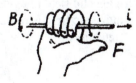

---

[*] For a negatively charged particle the force would point in the opposite direction.

[†] In the old days when some books defined current as the direction of electron flow, there was confusion between left-hand rules and right-hand rules. Today, fortunately, all textbooks define current as the direction in which positive charge would flow. So we safely use only right-hand rules for directions.

Faraday's law is a highlight of Chapter 25 in *Conceptual Physics* and is treated qualitatively. We imagine magnetic lines of force passing through the area of a wire loop. The product of the magnetic field and the area of the loop is called the *magnetic flux*, and is given the symbol Φ. As Faraday's law tells us, any change in the "amount" of magnetic flux through the area of a loop (or coil of loops) induces a voltage in that loop (or coil). This flux can be changed by

(1) increasing or decreasing the strength of the magnetic field B;
(2) changing the area of the loop or coil of loops in the field; or
(3) changing the loop's angular orientation to the field.

Figure 25.1 in your textbook illustrates the first case. Moving the bar magnet into or out of the coil strengthens or weakens the field passing through the coil (that is, the flux) and induces a voltage. Figure 25.8 show how the flux through a loop can change as the loop rotates in a uniform magnetic field. The relationship between induced voltage and changing flux is

$$\text{Voltage induced} \sim \text{number of loops} \times \frac{\Delta\,(\text{magnetic field}\times\text{loop area})}{\Delta\,\text{time}} \Rightarrow \mathcal{E}=-N\frac{\Delta\Phi}{\Delta t},$$

where $\mathcal{E}$ is the "electromotive force," (actually the induced voltage), N is the number of loops of wire in the coil, and $\dfrac{\Delta\Phi}{\Delta t}$ is the rate of change of the magnetic flux.*

The negative sign arises because experiments show that an induced voltage always gives rise to a current whose magnetic field opposes the change in flux. This is known as Lenz's law. If Lenz's law were not true, an induced current could produce a flux in the same direction as the change and grow in an ever-increasing crescendo—a conservation of energy no-no! Lenz's law is consistent with the conservation of energy.

Sample Problem 1

A particle of mass m with a charge q moves to the right at velocity v in a uniform magnetic field B. Velocity v and field B are at right angles to each other.
(a) What is the magnitude of the force that the magnetic field exerts on the particle?

Answer:
$F = qvB$, a one-step solution.

(b) The magnetic field B points in a direction out of the page. In what direction will the particle be deflected by the magnetic field?

Direction of F = ?
A convention for showing the direction of the magnetic field imagines that the direction of the magnetic field is represented by an arrow with fletchings:

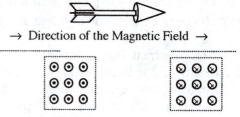

$\rightarrow$ Direction of the Magnetic Field $\rightarrow$

Representation of a
magnetic field pointing
out of the page

Representation of a
magnetic field pointing
into the page

---

* For a rotating coil the maximum magnetic flux is the product of the field strength B and the area of the coil through which the magnetic field lines pass ($\Phi = BA$), and occurs when the field is oriented perpendicular to the loop. As the coil steadily rotates, the flux changes in sinusoidal fashion from maximum to minimum and back again with each complete turn. A quantitative treatment of Faraday's law in generators where changing magnetic flux is due to rotation requires familiarity with radian units that are not covered in the textbook.

If the field were pointing out of the page you'd see just the circular head of the arrow and the point of its cone. If the field were pointing into the page, all you'd see is the feather fletchings at the rear.

We can determine the direction of the magnetic force using the right-hand rule. If you point your forefinger to the right (the direction of v) and your middle finger out of the page (the direction of B), your thumb points down the page (the direction of F). The particle will be deflected downward.

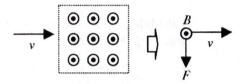

Had q been a negative charge moving in the same direction, the force would have been in the opposite direction, upward.

(c) Because the magnetic force on a charged particle is always perpendicular to the velocity, the trajectory of the charged particle in the magnetic field region will be a circular arc. What is the radius of this arc? (Recall that any particle moving in a circular path is acted on by a centripetal force.)

Focus: $r = ?$

The magnetic force acts as the centripetal force on the particle.

From $qvB = \dfrac{mv^2}{r} \Rightarrow r = \dfrac{mv}{qB}$. This answer makes sense. A stronger magnetic field or a more highly charged particle will result in a large magnetic force, pulling the particle into a tighter circle. A more massive or faster particle will have a greater tendency to move in a straight line, so it would describe a circle with a larger radius.

(d) Calculate the radius of the circular arc if the mass of the particle is 1.5×10^{-9} kg, its charge is 5.0×10^{-5} C, its velocity is 1.2×10^{3} m/s, and the magnetic field is 0.18 T.

Solution: $r = \dfrac{mv}{qB} = \dfrac{(1.5\times10^{-9}\,\text{kg})(1.2\times10^{3}\,\text{m/s})}{(5.0\times10^{-5}\,\text{C})(0.18\ \text{T})} = \mathbf{0.20\ m.}$

Sample Problem 2
The force on a charged particle moving perpendicularly in a magnetic field is given by $F = qvB$.
(a) Show that for an electric current, the equation for magnetic force becomes $F = ILB$, where I is the current, L is the length of wire in and perpendicular to the magnetic field, and B is the magnetic field strength.

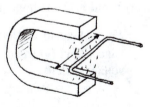

$F = ?$

From $F = qvB$, where $v = \dfrac{\text{distance } L}{\text{time } t} = \dfrac{L}{t} \Rightarrow F = q\left(\dfrac{L}{t}\right)B = \dfrac{q}{t}LB = \mathbf{ILB}$ (where $\dfrac{q}{t} = I$).

(b) How large is the magnetic force on a length of wire L that carries a current I and is placed at right angles to a field of strength B?

Answer: The equation for force as found in part (a), $F = \mathbf{ILB}$.

(c) **Calculate the force on a wire of length 0.10 m carrying a 2.0-A current in a 0.0030-T magnetic field.**

Solution: $F = ILB = (2.0\,\text{A})(0.10\,\text{m})(0.0030\,\text{T}) = \mathbf{0.00060\,N}$.

(d) **Calculate the force on 100 strands of the same wire, each with the same current in the same magnetic field.**

Solution: The magnetic force acts on each of the 100 strands, so the total force on the stranded wire is 100 times greater than the force on individual strand. $F = 100 \times ILB = 100 \times (2.0\,\text{A})(0.10\,\text{m})(0.0030\,\text{T}) = \mathbf{0.060\,N}$.

(e) **If the current is heading "clockwise" through the wire in the diagram, and the magnetic field points up from the lower pole to the upper pole of the magnet, in what direction is the force on the wire?**

Answer: The "positive charge carriers" are heading "into the page" through the poles of the magnet. B points up. The right-hand rule gives F pointing horizontally away from the magnet, **to the right**. So the wire tends to be ejected from the magnet.

(f) **When the direction of the current is reversed, how does this affect the direction of the force?**

Answer: Since the velocity vector for the charges reverses direction, the magnetic force acting on them also reverses direction. The wire tends to be pulled into the magnet, **to the left**.
A motor is a familiar example of a device in which the force can be alternated to and fro.

Sample Problem 3

A transformer (Figure 25.13 in the textbook) nicely employs Faraday's law. Consider an alternating voltage of 120 V that is applied to a step-up transformer having 140 turns on its primary and 1400 turns on its secondary.

(a) **What is the voltage in the secondary?**

Focus: Secondary Voltage = ?
Page 485 in *Conceptual Physics* shows the relationship between primary and secondary voltages relative to the numbers of turns:

$$\frac{\text{Primary voltage}}{\text{Number of primary turns}} = \frac{\text{secondary voltage}}{\text{number of secondary turns}}$$

Solving for secondary voltage,

$$\text{Secondary voltage} = \frac{\text{primary voltage} \times \text{number of secondary turns}}{\text{number of primary turns}} = \frac{(120\,\text{V})1400}{140} = \mathbf{1200\,V}.$$

This makes sense. The ratio of transformer turns is 10 to 1, so the voltage will be stepped up by a factor of ten.

(b) If the secondary coil is supplying a current of 0.1 A to a heater or other resistor, what is the current in the primary coil?

Answer: This transformer steps up the voltage from the primary coil to the secondary coil by a factor of ten. There is a corresponding step down of current in the secondary coil. The power into the transformer equals the power out, neglecting small losses. So $(IV)_{primary} = (IV)_{secondary}$, is consistent with the conservation of energy. Hence the current in the primary is ten times that in the secondary, or **1.0 A**.

(c) What is the power input to the transformer?

Focus: $P = ?$
From $P = IV$, we see that the power input is $(1.0 \text{ A})(120 \text{ V}) = $ **120 W**.

(d) By how much is the power stepped up in the secondary?

Answer: Whoa! Assuming no significant heat loss, the power in both coils is the *same*! Although voltages and currents can be stepped up or down in a transformer, energy, and therefore power, cannot. Ideally, the power in the secondary is also **120 W**. Conservation of energy rules!

Problems

25-1. A particle with charge $+q$ and velocity v is projected at right angles between the poles of a magnet of field strength B.
   (a) How large is the magnetic force on the particle in the magnetic field?
   (b) In what direction is the magnetic force on the charged particle?
   (c) Calculate the magnitude and direction of the force on the particle if its charge is 0.0020 C, its velocity is 550 m/s, and the magnetic field between the pole pieces is 0.0020 T.
   (d) If the charge were instead $-q$, what would be the difference in the direction of the force?

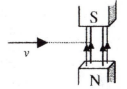

25-2. A beam of electrons traveling to the right at speed v is directed perpendicularly into a uniform magnetic field of strength B pointing into the page.
   (a) How large is the magnetic force on each of the electrons?
   (b) Calculate the deflecting force on each electron in the beam if they travel at 3000 m/s and the strength of the magnetic field is 0.15 T.
   (c) In what initial direction will the electrons be deflected?

25-3. Electrons traveling at speed v to the left into a uniform magnetic field experience a deflecting force F directed down the page.
   (a) Find the strength and direction of the magnetic field that produces the force F.
   (b) Calculate the strength of the magnetic field when the speed of the electrons is 2.0×10^7 m/s and the deflecting force is 3.0×10^{-13} N.

25-4. Mona directs a beam of protons perpendicularly into a magnetic field B. Each proton experiences a deflecting force F.
  (a) Find the speed of the protons.
  (b) Calculate the proton speed in a magnetic field of strength 0.080 T, where it experiences a deflecting force of 1.2×10^{-13} N.

25-5. A proton in a magnetic field travels at speed v in a circular path of radius r.
  (a) What is the strength of the magnetic field responsible for the circular path?
  (b) Calculate the magnetic field strength if the speed of the proton is 1.2×10^6 m/s and the radius of its curved path is 0.15 m.
  (c) Suppose that the magnetic field points into the page, and at one particular moment the proton is heading to the right. Will the proton follow a clockwise or a counterclockwise path? Defend your answer.

25-6. A proton in a magnetic field B travels in a circular path of radius r.
  (a) What is the speed of the proton?
  (b) What is the kinetic energy of the proton?
  (c) Calculate the kinetic energy if its path has a radius of 0.40 m and the magnetic field strength is 0.85 T.

25-7. A length of current-carrying wire L is placed at right angles in a magnetic field B.
  (a) What force is exerted on the wire when it carries a current I?
  (b) Calculate the force on the wire of length 0.15 m carrying a current 5.0 A in a field of 0.0050 T.
  (c) When the current heads counterclockwise between the poles of the magnet, the wire experiences a horizontal force into the magnet (toward the left). Which pole of the magnet is the north magnetic pole, the upper or the lower one?
  (d) If both the direction of the current and of the magnetic field are reversed, what effect, if any will this have on the direction of the force?

25-8. A length of current-carrying wire L placed at right angles to a magnetic field B experiences a force F.
  (a) What is the current in the wire?
  (b) Calculate the current in a wire of length 0.15 m in a field of 0.015 T when the force is 0.030 N.
  (c) If the direction of the current is reversed, what effect, if any, will this have on the direction of the force?

25-9. A horizontal rod of mass m, length L, and carrying a current I is aligned in a magnetic field for maximum magnetic force. The force is just right to keep the rod suspended in midair.
  (a) What is the direction of current through the rod, clockwise or counterclockwise? Defend your answer.
  (b) Find the magnetic field strength B.

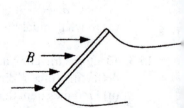

25-10. A horizontal rod of length L carrying a current I, is suspended at right angles into a horizontal magnetic field (as shown in the sketch of the previous problem). The strength of the field is B. The magnetic force on the rod is measured by the balance and is found to be F.
  (a) What is the current in the rod?
  (b) Calculate the current if the force on the rod is 0.15 N, its length is 0.20 m, and the magnetic field is 0.055 T.
  (c) If the magnetic force on the rod is downward, in what direction is the current running in the circuit? (Clockwise or counterclockwise?)

25-11. Recall that the force on a charged particle in an electric field is qE. The force on a charge $+q$ moving at right angles to a magnetic field is qvB.
  (a) In the diagram to the right, what is the direction of the electric force that will act on the particle?
  (b) What is the direction of the magnetic force that will act on the particle?
  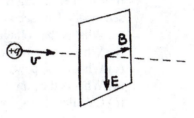
  (c) If adjustments are made so that the electric and magnetic forces on a moving charged particle are equal and opposite, the particle will be undeflected. Show that in this situation the speed of the particle is given by $v = E/B$.
  (d) If the electric field is supplied by equal and oppositely charged parallel plates, where the electric field is $E = V/d$, show that a more practical version of the above equation is $v = V/(Bd)$.
  (e) In a mass spectrometer (Figure 34.14 on page 671 in *Conceptual Physics*) it is important that all ions have the same velocity before entering the magnetic field that will deflect them according to mass. Hence a "velocity selector" for ions is needed. How do the above equations provide the foundation for a device that selects only particular velocities of charged particles?

25-12. An electric field E exerts a force F on an ion of charge q. At right angles to the electric field is a magnetic field B.
  (a) What is the speed of the ion if the electric and magnetic forces are equal and opposite?
  (b) Calculate the speed if the charge of the ion is 1.6×10^{-19} C, the electric field is 6.0×10^{6} N/C, and the magnetic field is 0.83 T.
  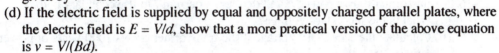
  (c) If the charge were twice as great, what strength of magnetic field would result in a straight-line path?

25-13. A charged-particle q with initial velocity v enters a region of uniform electric and magnetic fields perpendicular to each other. The particle's path is perpendicular to the magnetic field of strength B.
  (a) What should be the strength of the electric field if the particle is to pass through the fields undeflected?
  (b) Calculate the electric field strength if the charge of the particle is 1.6×10^{-18} C, its velocity is 5.5×10^{4} m/s and the strength of the magnetic field is 1.35 T.

25-14. After passing through a velocity selector, a stream of ions
moving at velocity v enters a uniform magnetic field B and is
swept into a circular arc of radius r.
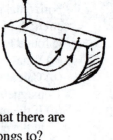
(a) What is the mass of each ion?
(b) Calculate the mass of the ion if its charge is 1.6×10^{-19} C, its
speed in the field is 2.05×10^5 m/s, the radius of its path is
0.308 m, and the magnetic field strength is 1.36 T.
(c) Calculate the mass of the ion in atomic mass units (amu), given that there are
6.02×10^{26} amu in 1 kg. Can you guess what element this ion belongs to?

25-15. An alternating voltage V is applied to the primary coil of a step-up
transformer having N turns on its primary and $20\,N$ turns on its
secondary. The secondary current is I.
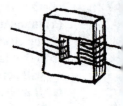
(a) What is the voltage in the secondary?
(b) What is the primary current?
(c) What is the power input?
(d) What is the power output, assuming no losses?
(e) Calculate the answers to the above questions for a primary 120 V and a secondary
current 0.1 A.

25-16. A transformer draws power P and delivers a current I in its secondary coil.
(a) What is the secondary voltage?
(b) Calculate the secondary voltage if the transformer draws 1000 watts of power, and
delivers 25 A to its secondary.
(c) What is the ratio of primary to secondary turns if the primary voltage is 120 V?

25-17. Roy uses a step-down transformer to power a toy electric train.
(a) If the primary coil of the transformer has N turns, and the
desired output is 6 volts from a household circuit of 120 V,
how many turns should be on the secondary?
(b) Calculate the number of turns on the secondary if the
primary has 240 turns.
(c) Why would it be a bad idea to reverse the input and output of the transformer; that is,
to apply 120 volts to the secondary?

25-18. The uniform magnetic field between the poles of an
electromagnet increases at a steady rate and induces an emf $\mathcal{E}$
in a single conducting loop of wire placed in the field.
(a) What is the induced current in the loop if an emf $\mathcal{E}$ is
induced and the resistance of the loop is R?
(b) Calculate the induced current for an induced emf of 4.0 V
and a loop resistance of 0.8 ohms.
(c) Does induced voltage depend on circuit resistance? Does induced *current* depend on
circuit resistance? Defend your answers.

25-19. A coil of wire containing N circular loops, each of cross-sectional area A, is placed between the pole pieces of a magnet where the field is uniform and of field strength B. The area of the loops of the coil are at right angles to the field. The field then decreases at a steady rate to zero in time t.
  (a) What is the magnitude of the induced voltage in the coil?
  (b) Calculate the magnitude of the induced voltage if the field decreases from 0.36 T to zero in 2.0 s and the coil has 400 loops of cross-sectional area 0.050 m^2.

25-20. A coil of wire with N turns and an area A is placed in a magnetic field B and oriented such that the area is perpendicular to the field. The coil is then rotated a quarter-turn, 90°, in time t so it ends up parallel to the field.
  (a) What is the magnitude of the average induced voltage in the coil?
  (b) Calculate the average induced voltage if the magnetic field is 1.2 T, the number of turns in the coil is 20, the area of the loops is 0.045 m^2, and the coil is flipped a quarter-turn in 0.45 s.

Show-That Problems

25-21. An electron is projected at 3.0×10^6 m/s into a 0.13-T magnetic field.
  Show that the maximum deflecting force on the electron is 6.2×10^{-14} N.

25-22. An electron is projected into a magnetic field of strength 0.15 T and undergoes a deflecting force of 1.50×10^{-14} N.
  Show that the speed of the electron is 6.3×10^5 m/s.

25-23. A particle with a velocity of 1.0×10^4 m/s and charge 0.00010 C encounters a perpendicular magnetic field and is deflected with a force of 3.5×10^2 N.
  Show that the strength of the magnetic field is 3.5×10^{-2} T.

25-24. A vertical beam of particles with a mass 12 times that of a proton and a charge of 9.6×10^{-19} C enters a magnetic field of 0.30 T and is bent in a semicircle of radius 0.50 m.
  Show that the speed of the particles is 7.2×10^6 m/s.

25-25. A particle of mass 0.00020 kg and charge 0.00010 C is moving at 3,000 m/s. It encounters a perpendicular magnetic field of strength 0.44 T.
  Show that it is swept into a curved path with a radius slightly greater than 13 km.

25-26. A particle moving at 600 m/s with charge of 20 μC enters a magnetic field of 1.6 T and is swept into a circular arc of radius 225 m.
  Show that the mass of the particle is 1.2×10^{-5} kg.

25-27. A current of 10.0 A flows through a rod of length 0.20 m placed perpendicularly across a magnetic field of 0.050 T.
  Show that the magnetic force on the rod is 0.10 N.

25-28. A force of 0.60 N acts on an 8.00-A current-carrying rod of length 0.300 m placed at right angles to a magnetic field.
  Show that the strength of the magnetic field is 0.25 T.

25-29. A step-up transformer boosts 12 V to 120 V.
Show that the current in the secondary is one-tenth as much as the current in the primary.

25-30. The secondary coil of an ideal transformer has 600 turns and the primary has 30 turns.
Show that when the primary is connected to a 120-V line, the output of the secondary will be 2400 V.

25-31. The secondary coil of an ideal transformer has 600 turns and the primary has 30 turns.
Show that when connected to a 120-V line with a current of 20 A in the primary, the current in the secondary will be 1 A.

25-32. The transformer on a utility pole steps the voltage down from 20,000 V to 220 V for use in a home.
Show that the ratio of primary to secondary turns in the transformer is 91 to 1.

25-33. In the transformer of the previous problem, the home uses 2.0 kW of power.
Show that the primary and secondary currents in the transformer are 0.1 A and 9.0 A, respectively.

25-34. A velocity selector is designed to pass protons through undeflected only if they have a speed of 5.0×10^5 m/s.
Show that if the magnetic field used is 0.40 T, then the electric field oriented perpendicular to B must have a strength of 2.0×10^5 N/C.

25-35. A circular loop with an area of 0.012 m^2 is in a uniform magnetic field of 0.25 T.
Show that the flux through the loop when the field is oriented perpendicular to the loop is 0.0030 T·m^2.

25-36. A circular loop of radius 0.30 m is positioned in a magnetic field of 0.18 T.
Show that the maximum flux in the loop is 0.051 T·m^2.

25-37. Suppose the loop in the previous problem is in a magnetic field that decreases from 0.18 T to zero in 0.50 s.
Show that the induced voltage in the loop is 0.10 V.

25-38. Suppose that an ammeter is connected to the loop of the previous problem. The total resistance of the loop and the meter is 0.5 ohms.
Show that the current pulse is 0.2 A.

25-39. A coil of 100 turns of square loops of wire with an area of 0.50 m^2 is in a region where a magnetic field increases by 0.30 T in 0.10 s.
Show that the induced voltage in the coil is 150 volts.

25-40. Suppose that an ammeter is connected to the coil of the previous problem. The total resistance of the coil and the meter is 5.5 ohms.
Show that the current is about 27 A.

28 Reflection and Refraction

In this chapter we confine treatment of reflection to plane mirrors. Although mirrors with curved surfaces are briefly mentioned in the textbook, that coverage is not extended here. Light doesn't change speed upon reflection, but its speed does change when it undergoes refraction. We extend refraction to simple lenses. We make use of Snell's Law and the thin-lens equation, which are treated only in footnotes in the textbook.

The Law of Reflection

When light is incident at an angle on a reflecting surface it is reflected at the same angle. This is the law of reflection:

Angle of incidence = angle of reflection

It is customary to measure the angles from the normal to the reflecting surface as shown.

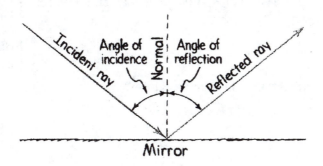

Sample Problem 1

A beam of light is incident on a plane mirror at an angle θ relative to the normal.

(a) What is the angle between the reflected ray and the normal?

> *Answer*: In accord with the law of reflection, the angle of reflection equals the angle of incidence θ, from the other side of the normal.

(b) What is the angle between the reflected ray and the mirror *surface*?

> *Answer*: The angle with respect to the mirror surface is the complement of θ, which is **$90° - \theta$**. For example, a ray at an angle of 30° to the normal would make an angle of 90°–30° or 60° with the surface. To avoid confusion, angles are always expressed relative to the normal in optical situations.

Refraction

When light bends in passing obliquely from one medium to another we call the process *refraction*. Refraction is a consequence of light traveling at different speeds in different media. It is customary to compare the speed of light in a vacuum, c, with the speed of light in other media. In so doing we define a ratio called the *index of refraction, n*.

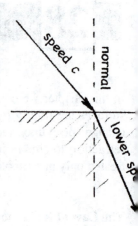

$$n = \frac{c}{v} = \frac{\text{speed of light in a vacuum}}{\text{speed of light in a medium}}$$

For example, light in a diamond travels at 124,000 km/s, and so for diamond $n = \dfrac{300,000 \text{ km/s}}{124,000 \text{ km/s}} = 2.42.$

The indices (plural for index) of refraction shown in the table are given for a specific wavelength of light because different frequencies of light have different speeds in a transparent medium. Note that n is always greater than 1 because the speed of light in a vacuum is greater than the speed of light in any medium. Since n is a ratio of speeds it has no units. The values of n in the table may be useful for the problems that follow.

Indices of Refraction (at $f = 5.0 \times 10^{14}$ Hz)	
Air	1.00029
Water	1.33
Ethyl alcohol	1.36
Fused quartz	1.46
Oil (typical)	1.50
Glass (by type)	1.45-1.7
Diamond	2.42

Sample Problem 2

What is the index of refraction of light in a vacuum?

Answer: $n = \dfrac{v}{c} = \dfrac{c}{c} = 1.$

Snell's Law

In about 1621, just after Pilgrims had landed at Plymouth Rock in America, the Dutch astronomer and mathematician Willebrord Snell discovered a relationship between the indices of refraction of two media and the angles of incidence and refraction as the light passes from one medium into the other. Snell's law for light passing from medium 1 to medium 2 is:

$$n_1 \sin \theta_1 = n_2 \sin \theta_2$$

where n_1 and n_2 are the respective indices of refraction of the media in which the light travels. As always, angles θ are taken with respect to the normal. The same law holds if the light is going in the other direction, from medium 2 to medium 1.

Snell's law can be written to show the relationship between the speed of light in each medium and the angles with the normal that an oblique ray makes as it passes from one medium to another.

$$\frac{\sin \theta_1}{\sin \theta_2} = \frac{v_1}{v_2}$$

Sample Problem 3
A merman beneath the surface of the sea looks above at a woman standing on a rock. Light from her hair on the way to his eye is incident upon the water at an angle θ_1 to the normal, and is refracted at the surface to his eyes. The index of refraction of water is n_w.

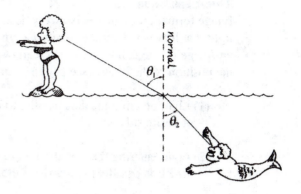

(a) What is the angle of refraction θ_2 for light rays from her hair heading toward the merman's eyes?

> *Focus*: Angle of refraction $\theta_2 = ?$
> Snell's law provides the necessary physics.
> From $n_1 \sin\theta_1 = n_2 \sin\theta_2$ $\Rightarrow \sin\theta_2 = \frac{n_1}{n_2}\sin\theta_1$. So θ_2 is the angle whose sine is $\frac{n_1}{n_2}\sin\theta_1$. We
> can write this as $\theta_2 = \arcsin\left(\frac{n_1}{n_2}\sin\theta_1\right)$, or equivalently, $\theta_2 = \sin^{-1}\left(\frac{n_1}{n_2}\sin\theta_1\right)$. Here we have
> $n_1 = 1$ for air, and n_2 is n_w. So $\boldsymbol{\theta_2 = \sin^{-1}\left(\frac{1}{n_w}\sin\theta_1\right)}$.

(b) Calculate the angle of refraction given that the angle of incidence is 60°. The index of refraction for water is given in the table above.

> *Solution*: $\theta_2 = \sin^{-1}\left(\frac{1}{n_w}\sin\theta_1\right) = \sin^{-1}\left(\frac{1}{1.33}\sin 60°\right) = \sin^{-1}\left(\frac{0.866}{1.33}\right) = \sin^{-1} 0.65$. Depending on
> your type of calculator, try entering 0.65, and then tap the inverse sine function.
> You should find the angle to be **41°**. (If not, get help from your instructor.)

(c) If the merman shone a laser pen along his line of sight to the woman, would it shine on her hair?

> *Answer*: Yes. The path of light in one direction is the same path in the opposite direction, a
> rule that applies to all optical systems, however complicated. It is called the
> *principle of reversibility*.

(d) If the merman pushed a long straight stick along his line of sight to the woman, would it reach her?

> *Answer*: No. Whereas the path of light changes where the water and air meet, the stick would
> remain straight and therefore pass above the woman's head. Interestingly, the stick
> would look bent to the merman.

Thin-Lens Equation

Image formation by lenses is nicely described using ray diagrams. In the diagram below, all of the light that leaves the candle (the *object* in this case) and passes through the lens converges to form an *image* of the candle on the screen. Any two of three principal rays are sufficient for locating the position of the image (see page 115 in the *Practicing Physics* book).

(1) Light entering the lens parallel to its axis is refracted through the focal point on the opposite side.

(2) Light entering the lens through the near focal point is refracted through the lens in a direction parallel to the axis of the lens.

(3) Light entering the center of the lens passes through undeflected.

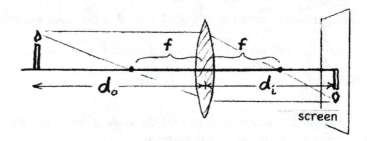

The sketch above shows two of these principle rays. (Sketch the third through the center of the lens for a check.) A quantitative way of relating object distances with image distances is given by the thin-lens equation:

$$\frac{1}{d_o} + \frac{1}{d_i} = \frac{1}{f} \quad \text{or} \quad d_i = \frac{d_o f}{d_o - f}$$

Here d_o is the distance of the object from the lens, d_i is the distance from the lens to the image, and f is the focal length of the lens. Distances are positive (+) when object and image are on opposite sides of the lens. When the image is on the same side of the lens as the object, d_i is negative (−) (this situation is illustrated for a convex lens in figure 28.48 on page 549 of the *Conceptual Physics* textbook). The magnification of a lens, the height of the image h_i compared with the height of the object h_o, is given by

$$M = \frac{h_i}{h_o} = -\frac{d_i}{d_o}$$

A negative value of M means that the image is inverted. For a negative image distance d_i (image on the same side of the lens as the object) magnification is positive and the image is right-side up.

Lenses that converge parallel rays of light are convex and have positive (+) focal lengths. Lenses that diverge parallel rays of light are concave and have negative (−) focal lengths.

Sample Problem 4

A college student amazes her little brother with a demonstration of real images. She uses a strong magnifying glass to project an image of a candle onto a piece of paper several centimeters away. The lens has a focal length of 3.0 cm. The flame of the candle is about 1.0 cm tall and is 5.0 cm in front of the lens.

(a) **Predict the position of the image and its approximate size by drawing a ray diagram showing three rays from the candle flame.**

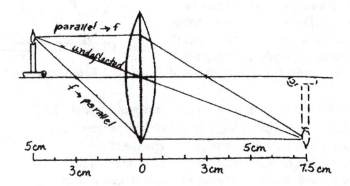

Analysis: The image position is located at the intersection of three rays: One parallel to the axis, that after refracting through the lens goes downward through the opposite-side focal point; another that goes straight through the center of the lens without deflection; and a third ray that passes through the near focal point and exits the lens parallel to the axis.

(b) **Use the thin-lens equation to calculate the distance between the lens and the image.**

Solution: $d_i = \dfrac{d_o f}{d_o - f} = \dfrac{(5.0 \text{ cm})(3.0 \text{ cm})}{(5.0 \text{ cm} - 3.0 \text{ cm})} = \textbf{7.5 cm}.$

(c) **How large will the image be?**

Solution: $h_i = ?$ From $M = \dfrac{h_i}{h_o} = -\dfrac{d_i}{d_o} \Rightarrow h_i = \left(-\dfrac{d_i}{d_o} \right) h_o = \left(-\dfrac{7.5 \text{ cm}}{5.0 \text{ cm}} \right) 1.0 \text{ cm} = \textbf{-1.5 cm}.$

The image is 1.5 cm tall. The minus sign tells us that the image is inverted.

Problems

28-1. A beam of light is incident on a plane mirror at an angle θ relative to the normal.
  (a) What is the angle between the reflected ray and the normal?
  (b) What is the angle between the incident and reflected rays?

28-2. A light beam reflects to and fro between ordinary parallel mirrors as shown.
  (a) Sketch the path of the beam as it reflects between the mirrors and estimate the number of reflections that occur before the beam is outside the mirror system?
  (b) How will the strength of the beam exiting the mirror system compare with its strength before entering?

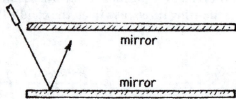

28-3. • A beam of light is incident upon a plane mirror at angle θ. The mirror is then rotated a bit by angle α as indicated in the sketch.
  (a) By how much is the angle of reflection changed?
  (b) A beam of light is incident on a mirror at an angle of 30° with the normal, and reflects onto a vertical screen 10.0 m away. If you rotate the mirror by 2° how far will the spot of light on the screen move?

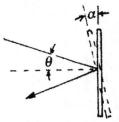

28-4. You walk toward a mirror at speed v.
  (a) How fast do you and your image approach each other?
  (b) In walking away from the mirror at speed v, how fast do you and your image recede from each other?

28-5. A pool table has dimensions $W \times L$. A pool ball strikes the cushion at an angle θ to the normal as shown.
  (a) Assuming the ball obeys the law of reflection, at what angle does it bounce from the cushion?
  (b) How far does it travel before it strikes the opposite cushion?
  (c) Under what conditions will the ball not obey the law of reflection? (Hint: What does it mean to put "English" on the ball?)

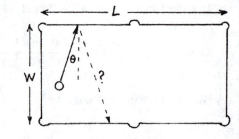

28-6. A beam of light is incident upon a pane of window glass at 60° to the normal.
  (a) What is the angle that the refracted beam makes with the normal inside the glass if the index of refraction of the glass is 1.5?
  (b) At what angle will the beam emerge from the glass? (Assume that the glass surfaces are parallel to each other.)

28-7. Light travels at different speeds in glass and in a diamond.
(a) Find the ratio of the speed of light in glass ($n = 1.50$) to the speed of light in diamond ($n = 2.42$).
(b) Once light is inside a diamond it may reflect from an inner surface. How does the angle of reflection inside the diamond compare with the angle of incidence inside the diamond?

28-8. Flint glass has an index of refraction $n = 1.638$ for red light and $n = 1.675$ for blue light.
(a) How much faster (in m/s) does red light travel in this glass compared with blue light?
(b) In air, red light takes 1.00×10^{-10} s to cover 3.00 cm. How long does it take to travel 3.00 cm through flint glass?

28-9. Consider the same flint glass of the previous problem. A beam of white light shines through air onto it at an angle θ.
(a) Calculate the angle of refraction for the blue light when the incident beam is 36.00° to the normal.
(b) Calculate the angle of refraction for the red light in the same beam.

28-10. Rainbows are formed when white light from the Sun meets raindrops in the atmosphere. Water has an index of refraction of $n = 1.3311$ for red light and $n = 1.3330$ for yellow light.
(a) If the sunbeam shines on a raindrop at an angle of 41.00° with respect to the normal at a particular point of the drop, what is the angle of refraction in the drop for red light refracted at that point?
(b) For yellow light?
(c) Light of both colors travels at very nearly the same speed in the air. Which one is slowed down more when it enters the drop? Defend your answer.

28-11. Looking straight downward, Marjorie notices that a coin at the bottom of a glass of water appears closer to her—closer by the index of refraction of air divided by the index of refraction of water.
(a) Calculate how far below the surface the coin appears in water 10.0 cm deep.
(b) In what sense does the coin appear magnified?

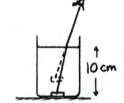

28-12. Experimental Ernie holds a converging lens in the sunlight over a piece of paper and the paper ignites.
(a) What is the likely focal length of the lens if the lens is held 15 cm above the paper?
(b) Why would the paper not ignite at other distances from the lens?

28-13. When an object is located a distance x to the left of a lens, the image is formed at a distance $0.5x$ to the right of the lens.
(a) What is the focal length of the lens?
(b) If the object was instead $2x$ to the left of the lens, where would the image appear?

28-14. An object of height h is placed at distance x to the left of a converging lens with a focal length f.

  (a) What is the image distance?

  (b) What is the height of the image?

  (c) Calculate values for (a) and (b) for a 5.0-cm tall object located a distance 7.0 cm to the left of a lens of focal length 15 cm. (See Figure 28.48 on page 549 in *Conceptual Physics* for a ray diagram for this situation.)

  (d) Calculate the values if instead we used a diverging lens of focal length –15 cm.

28-15. A diverging lens is used to view an object placed on the opposite side of the lens.

  (a) If the focal length of the lens is –33 cm, find the image distance when an object is located 25 cm from the lens.

  (b) What is the magnification at this distance?

  (c) Will the image be right-side up or upside down? Defend your answer.

28-16. A lens for a 35-mm camera has a focal length 45 mm.

  (a) How close to the film should the lens be to form a sharp image of an object that is 10.0 m away?

  (b) What is the magnification of the image on the film?

28-17. A detective examines evidence with a standard converging magnifying glass. He holds the lens close enough to the evidence so that it produces an enlarged virtual image at his distance of most distinct vision.

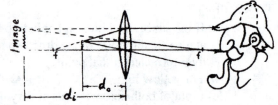

  (a) How far from the evidence should the detective hold the lens if its focal length is 15 cm, and the object position is adjusted so that the image distance is –22 cm (that is, on the same side of the lens as the object)?

  (b) What is the magnification?

  (c) What is the advantage of having the object located within the focal length?

28-18. When a physics instructor focuses on an exam paper, the lens in his or her eye forms a real image of the paper on the retina in the back of the eye. A wondrous series of physiological events occurs that turns the paper into a grade.

  (a) If the paper is 28 cm in front of the instructor's eye, and the lens of the eye is 3.0 cm in front of the retina, what is the focal length of the lens?

  (b) Is the eye lens a converging or a diverging lens?

28-19. The heat lamp in the bathroom has been absentmindedly left on. The light shines across the room, hitting a slender perfume bottle with convex sides that focus the light into a small dot that slowly but surely burns a hole in the wall behind the bottle.

  (a) If the heat lamp is 1.9 m from the bottle, which is 9.8 cm from the wall with the smoldering hole, what is the focal length of the bottle acting as a lens?

  (b) If the diameter of the bulb in the heat lamp is 14 cm, what is the diameter of its damaging image on the wall?

28-20. A jeweler's loupe is a small magnifying glass of short focal length that is held close to the eye to enable a very close look at a gem or watch part.

(a) If the loupe has a focal length of 2.2 cm and a gem being inspected is just far enough away from the lens so that the image distance is –22 cm from the lens (on the same side of the lens as the gem), how far from the lens is the gem being held?

(b) What will be the magnification?

Show-That Problems

28-21. A student 1.8 m tall stands in front of a plane mirror.
Show with a ray diagram that the smallest mirror to show her full image is 0.9-m tall. Further show that the distance from the mirror is not a factor.

28-22. A bee stands 0.3 m in front of a small mirror with a wall mirror 0.9 m behind him.
Show that the image of his stinger appears a distance 2.4 m in front of him.

28-23. Light slows when it enters a diamond, $n = 2.42$.
Show that the speed of light in a diamond is about 41% its speed in air.

28-24. Snell's law is introduced in this chapter in two ways.
Show how $\dfrac{\sin \theta_1}{\sin \theta_2} = \dfrac{v_1}{v_2}$ becomes $n_1 \sin \theta_1 = n_2 \sin \theta_2$.

28-25. A beam of light in air is incident upon a piece of crown glass at an angle of 37.0°. The index of refraction of crown glass is 1.52.
Show that the angle of refraction in the glass is 23.3°.

28-26. A laser beam in air is incident on a block of glass at an angle of 45°. The glass has an index of refraction of 1.50.
Show that the angle of refraction in the glass is 28°.

28-27. A laser beam beneath the surface of water is aimed upward toward the surface. Beyond a critical angle (where the light would refract along the water surface at 90° to the normal), light undergoes total internal reflection rather than refracting into the air above. Show that for water the critical angle is 48.8°.

28-28. The index of refraction for a diamond is 2.42.
Show that the critical angle in a diamond, beyond which internal reflection occurs, is 24.4°.

28-29. An object is placed 40.0 cm to the left of a lens and produces a real image at a distance of 80.0 cm to the right of the lens.
Show that the focal length of the lens is approximately 27 cm.

28-30. An object is placed 30 cm to the left of a converging lens and produces a real image at a distance of 15 cm to the right of the lens.
Show that the focal length of the lens is 10 cm.

28-31. A converging glass lens has an index of refraction of 1.60. Its focal length is 30.0 cm. Show that an object 2.5 m in front of the lens will appear in sharp focus when the screen is 34 cm in back of the lens.

28-32. An object is placed 16.0 cm in front of a converging lens with a focal length 10.0 cm. Show that the image distance is approximately 27 cm.

28-33. An object is placed 16 cm in front of a diverging lens with a focal length –10 cm. Show that the image distance is –6.25 cm.

28-34. A candle is placed 50.0 cm in front of a converging lens of focal length 10.0 cm. Show that the magnification of the candle image is – 0.25.

28-35. A converging lens with a focal length of 20.0 cm produces an image on a screen that is 2.0 m from the lens. Show that the object distance is 22 cm.

28-36. A converging lens is used to produce an image of an object. The object distance is twice the image distance. The object is 5.0 cm tall. Show that the height of its image is 2.5 cm tall.

28-37. An image that is 1.0 cm tall is formed on a screen behind a converging lens when an object 2.0 m tall is located 8.0 m in front of the lens. Show that the distance between the lens and the screen is 0.040 m.

28-38. A candle 10 cm tall is placed 30 cm from a thin converging lens. The crown glass lens has a focal length, f, of 20 cm, and an index of refraction, n, of 1.52. Show that the image of the candle is 60 cm from the lens.

28-39. A box 5.0 cm wide is placed 10.0 cm from a diverging lens. The image of the box, at a negative image distance, is one-fifth as wide as the box. Show that the focal length of the lens is –2.5 cm.

28-40. In biology lab you want to examine a piece of tissue at a magnification of +5.00. Show that if the tissue is 5.00 cm from the lens, then focal length of the lens is 6.25 cm.

33 | *The Atomic Nucleus and Radioactivity*

So far we've considered interactions between molecules (in solids, liquids, and gases, for example) and rearrangement of atoms within molecules such as occurs in chemical reactions. All of these interactions are fundamentally *electrical* in nature—they involve attractions and repulsions between the nuclei and the electrons surrounding them. Whatever changes occur leave the atoms themselves largely unchanged.

Nuclear reactions, on the other hand, involve additional forces, and usually lead to the formation of an entirely new nucleus. The electrical repulsion between the protons in the nucleus tends to push the nucleus apart, while the strong nuclear force acting between pairs of adjacent nucleons holds the nucleus together. Hence a nucleus needs a certain balance of neutrons and protons for stability—nearly equal numbers for light nuclei, and a preponderance of neutrons for heavy nuclei (where the mutual repulsion of the protons play an important role). If a nucleus is unstable because of "too many" protons, it will undergo a spontaneous change (a "decay") that *decreases* the ratio of protons to neutrons. If it is unstable because of "too many" neutrons, it will decay in such a way as to *increase* the ratio of protons to neutrons.

There are five basic ways that a nucleus can spontaneously change its identity. These changes constitute *radioactive decay*. The effects of different types of radioactive decay are summarized in the table below:

Type of decay	What happens:	Effect on the nucleus	Example:
Alpha decay (α)*	Nucleus ejects a helium nucleus (2 p, 2 n)	Nucleus has 2 less protons and 4 less nucleons overall	$^{235}_{92}U \rightarrow \, ^{231}_{90}Th + \, ^{4}_{2}\alpha$
Beta decay (β)*	Electron emitted, a neutron is changed into a proton[†]	Nucleus has one more proton, same number of nucleons as before	$^{239}_{92}U \rightarrow \, ^{239}_{93}Np + \, ^{0}_{-1}e$
Gamma decay (γ)	Nucleus "falls" from a higher to a lower state and emits a high-energy photon	Nucleus has the same number of protons and neutrons as before	$^{99}_{43}Tc^{*} \rightarrow \, ^{99}_{43}Tc + \, ^{0}_{0}\gamma$ (the * indicates that the nucleus is in an excited state).
Electron capture (EC)	Orbital electron captured, a proton is changed into a neutron*	Nucleus has one less proton, same number of nucleons as before	$^{40}_{19}K + \, ^{0}_{-1}e \rightarrow \, ^{40}_{18}Ar$
Positron emission (β⁺)	Positron emitted, a proton is changed into a neutron*	Nucleus has one less proton, same number of nucleons as before	$^{23}_{12}Mg \rightarrow \, ^{23}_{11}Na + \, ^{0}_{1}e$

If we are looking at the *atom* level, then we write $^{4}_{2}He$ rather than $^{4}_{2}\alpha$, since $^{235}_{92}U$ has 92 electrons, $^{231}_{90}Th$ has 90 electrons, and the alpha particle must pick up two electrons to become $^{4}_{2}He$ in order to balance out the charge.

Sometimes called beta-minus decay (β⁻) to distinguish it from the similar decay in which a positron is released (β⁺).

Another particle, called an *antineutrino*, is also formed. It has no charge, is very light (much less massive than an electron), and carries some of the momentum and energy from a nuclear decay. Neutrinos interact so weakly with matter that, for our purposes, we can ignore them.

A neutrino is also formed.

Radioactivity and half-life

Various degrees of "instability" exist among radioactive nuclei. Some radioactive samples will decay in less than a second; other samples will hardly decay at all in a person's lifetime. One way to characterize the degree of instability of a radioactive material is to report its *half-life*—the time for half of a radioactive sample to decay into something else. For instance, if you had 10 grams of a particular radioactive material, after one half-life only 5 grams of the original radioactive material would remain (the other 5 grams would have been converted to a different material). After two half-lives, 2.5 grams of the original radioactive material would remain. A material with a shorter half-life is less stable than one with a longer half-life—each nucleus within the less-stable sample has a higher probability of decaying in a unit time.

Suppose that you have a thousand dice and you roll them. Approximately one-sixth of them will come up as a one. If you remove all of the dice that come up as ones and roll the remaining dice again, the odds of any particular die coming up a one on the next roll hasn't changed—the odds are still one in six. But over time there will be fewer and fewer ones coming up because after each dice roll (and the removal of all of the dice that come up as one) there are fewer dice remaining to be rolled.

In this respect radioactive nuclei are like dice. If a certain nucleus doesn't decay in one time interval it has the same probability of undergoing decay in the next time interval. Over time, the number of decays from a given radioactive sample will decline, because the number of non-decayed nuclei remaining in the sample has declined.

When a material decays, most of the mass changes to another substance. Some is changed to energy in accord with the principle of mass-energy equivalence—next chapter.

Sample Problem 1

A bismuth-199 nucleus $\left(^{199}_{83}\text{Bi}\right)$ undergoes an alpha decay. What nucleus results?

Solution: In alpha decay the nucleus ejects an alpha particle—a helium nucleus composed of 2 protons and 2 neutrons. The overall positive charge in the nucleus is reduced by 2, and the overall number of nucleons is reduced by 4. Let X represent the unknown resulting nucleus.

Then $^{199}_{83}\text{Bi} \rightarrow {}^{199\text{-}4}_{83\text{-}2}\text{X} + {}^{4}_{2}\alpha$, where $^{195}_{81}\text{X}$ is the unknown. A numerical check confirms that both the number of nucleons and the overall charge are conserved. From the periodic table in your *Conceptual Physics* textbook (Figure 11.9 on page 219) you can see that the element with 81 protons is thallium, which has the symbol Tl. The new nucleus is $^{195}_{81}\text{Tl}$, thallium-195.

Sample Problem 2

A nucleus of barium-142, $^{142}_{56}\text{Ba}$, undergoes a beta-minus decay. What nucleus results?

Solution: In beta-minus decay a neutron within the nucleus turns into a proton and ejects an electron (the β^- particle) from the nucleus. This increases the positive charge in the nucleus by 1, while the overall number of particles in the nucleus remains the same.

So $^{142}_{56}\text{Ba} \rightarrow {}^{142}_{56+1}\text{X} + {}^{0}_{-1}\text{e}$. The new nucleus is $^{142}_{57}\text{La}$, lanthanum-142. Again, a check shows that both the number of nucleons and the overall charge are conserved.

ample Problem 3

hen a sample of boron-10 $\left(^{10}_5B\right)$ is bombarded with neutrons the nucleus can be ansmuted to a new nucleus by first absorbing a neutron, and then decaying into a new ement by alpha decay. What nucleus is ultimately formed?

Solution: The reaction is $^{10}_5B + ^1_0n \rightarrow$? $\rightarrow$? $+ ^4_2\alpha$.

The first question mark is $^{10}_5B + ^1_0n \rightarrow ^{11}_5B$ since the nucleus adds a nucleon without changing its charge. The alpha decay takes away two protons and a total four nucleons from the boron-11 nucleus, so $^{11}_5B \rightarrow ^7_3Li + ^4_2\alpha$. The final nucleus formed is ^7_3Li.

ample Problem 4

atch dials are often painted with a compound containing tritium, ^3_1H, which has a half- fe of 12.3 years. When the tritium atom undergoes beta decay the emitted electron strikes fluorescent pigment, which then glows.

) If your tritium-dial watch is giving off a certain amount of light, how many half-lives must pass until the watch gives off only one-eighth as much light?

Focus: $t = ?$

When the watch dial produces one-eighth as much light as initially, there must be one-eighth as many beta particles being emitted and striking the fluorescent pigments on the dial. So one-eighth of the original number of tritium atoms must still be present.

During each half-life the number of tritium nuclei in the sample (and the number of particles being emitted by the sample) is reduced by half:

Number of half- lives that have occurred	Fraction of the original sample remaining	Fraction of original number of particles being emitted
1	1/2	1/2
2	1/4	1/4
3	1/8	1/8
4	1/16	1/16

It will take **3 half-lives**, or 3×12.3 yr ≈ 37 yrs for the watch to dim to one-eight of its original brightness.

) What nucleus is formed when the tritium undergoes β decay?

Answer: In beta decay a neutron within the nucleus becomes a proton as an electron is ejected. The number of protons in the nucleus increases by 1 while the total number of nucleons remains the same.

So $^3_1H \rightarrow ^?_?X + ^0_{-1}e \Rightarrow ^?_?X = ^3_2He$, an isotope of helium.

) Why do the beta rays emitted by the face of the watch pose no health hazard?

Answer: The emitted electrons don't have enough energy to penetrate the watch case or crystal (see Figure 33.4 in the textbook) and hence pose no health hazard.

Sample Problem 5

Uranium-235 dating is useful in estimating the age of ancient rocks. Uranium-235, through a series of decays, transforms to lead-207. The half-life for this decay process is 704 million years. By measuring the ratio of lead-207 to uranium-235 in the rock one can determine how long ago the rock was formed, assuming that the rock initially contained no lead-207.
a) Complete the table below:

Number of half-lives that have passed since the rock formed	Ratio of lead-207 atoms to uranium-235 atoms is
1 half-life	1:1
2 half-lives	?
3 half-lives	?

Solution: Let's let N_0 represent the original number of uranium-235 atoms in the sample. At the end of one half-life, half of the uranium-235 atoms have decayed to produce lead-207. At that time the ratio lead-207 atoms : uranium-235 atoms is

$$\tfrac{1}{2}N_0 : \tfrac{1}{2}N_0 \Rightarrow \text{a } 1:1 \text{ ratio.}$$

After two half-lives, half of the remaining uranium-235 atoms have decayed into lead-207 and only one-quarter of the original uranium-235 atoms remain. The ratio lead-207 atoms : uranium-235 atoms is

$$\tfrac{3}{4}N_0 : \tfrac{1}{4}N_0 \Rightarrow \text{a } 3:1 \text{ ratio.}$$

After three half-lives, half of the remaining uranium-235 atoms have decayed, so only one-eighth of the original uranium-235 atoms remain. Seven-eighths have been transformed to lead-207. The ratio lead-207 atoms: uranium-235 atoms is then

$$\tfrac{7}{8}N_0 : \tfrac{1}{8}N_0 \Rightarrow \text{a } 7:1 \text{ ratio.}$$

The completed table looks as follows:

Number of half-lives that have passed since the rock formed	Ratio of lead-207 atoms to uranium-235 atoms is
1 half-life	1:1
2 half-lives	3:1
3 half-lives	7:1

(b) A geologist finds an ancient volcanic rock where the ratio of lead-207 atoms to uranium-235 atoms is 2.9:1. Approximately how old is the rock?

Answer: We determined that if the ratio were 3:1 the rock would be two half-lives old, which is $= 2 \times (704 \text{ million yr}) = 1.408$ billion years. Since the ratio of lead-207 to uranium-235 atoms is almost, but not quite 3:1 we'd estimate the rock to be just a bit younger, approximately **1.4 billion years old**.

33-1. What nucleus results when $^{81}_{36}\text{Kr}$ undergoes electron capture?

33-2. What nucleus results when $^{27}_{14}\text{Si}$ undergoes positron emission?

33-3. Complete the following table:

	Initial nucleus	Decay mode	Equation describing the decay	Resulting nucleus
(a)	$^{232}_{90}\text{Th}$	α		
(b)	$^{140}_{56}\text{Ba}$	β^-		
(c)	$^{51}_{25}\text{Mn}$	β^+		
(d)	$^{109}_{49}\text{In}$	EC		
(e)	$^{60}_{27}\text{Co}^*$	γ		
(f)	$^{74}_{33}\text{As}$	β^+		
(g)	$^{3}_{1}\text{H}$	β^-		
(h)	$^{239}_{94}\text{Pu}$	α		
(i)	$^{240}_{96}\text{Cm}$	α		
(j)	$^{95}_{40}\text{Zr}$	β^-		
(k)	$^{210}_{85}\text{At}$	EC		

33-4. A uranium-238 nucleus $\left(^{238}_{92}\text{U}\right)$ in a nuclear reactor can absorb a neutron and then experience two beta decays. What nucleus is formed?

33-5. A nitrogen-14 nucleus $\left(^{14}_{7}\text{N}\right)$ can absorb a high-energy neutron resulting from interactions between cosmic rays and the atmosphere, and then eject a proton from its nucleus to form the nucleus of a new element. What element is formed?

33-6. When oxygen-16 $\left(^{16}_{8}\text{O}\right)$ absorbs a neutron and then undergoes an alpha decay, it forms a nucleus that is useful for studying chemical reactions in living organisms. What nucleus is formed?

33-7. Nuclear reactors are sometimes started up with neutrons originating from a beryllium-9 $\left(^{9}_{4}\text{Be}\right)$ target that has been bombarded by alpha particles. This produces a new nucleus and an ejected neutron. What is the identity of the new nucleus?

33-8. Americium-241 $\left(^{241}_{96}\text{Am}\right)$ is used in smoke detectors. It undergoes an alpha decay with a half-life of 432 years. The alpha particles ionize some of the air near the americium-241 source, and these charged ions are collected on a metal plate that has a small voltage applied to it. The collected particles cause a small electric current to flow in the detector. If smoke particles get in between the americium-241 source and the detector and reduce the current, the alarm is triggered.

(a) Americium-241 is formed in a nuclear reactor when plutonium-239 $\left(^{239}_{94}\text{Pu}\right)$ absorbs two neutrons and then undergoes beta decay. Write the nuclear reaction for the formation of americium-241.

(b) A typical smoke detector experiences 370,000 alpha decays each second. How much time is needed for the number of decays to reduce to 93,000 decays per second?

33-9. Plutonium-238 $\left(^{238}_{94}\text{Pu}\right)$ is useful as a power source for heart pacemakers and for spacecraft. Heat from its nuclear decay is used by specialized semiconductor chips to produce electric current in a device called a Radioisotope Thermoelectric Generator (RTG). Plutonium-238 is manufactured in nuclear reactors by bombarding neptunium-237 $\left(^{237}_{93}\text{Np}\right)$ with neutrons. The neptunium-237 nucleus absorbs a neutron and then undergoes beta decay to form the plutonium-238.

(a) Write the nuclear reaction for the formation of plutonium-238.

(b) Plutonium-238 has a half-life of about 88 years. How much time would have to elapse for the electric current in your RTG to fall to $\frac{1}{16}$ of its original value?

(c) Would you have to worry about replacing the power source of a plutonium-238 powered pacemaker? Why or why not?

33-10. Strontium-90 $\left(^{90}_{38}\text{Sr}\right)$ undergoes beta-minus decay with a half-life of about 29 years. The Russian navy used it to power remote navigational beacons and lighthouses. A strontium-90 based Radioisotope Thermoelectric Generator (RTG) can produce up to 128 watts of electrical power from the heat of 9.6×10^{15} radioactive decays per second.

(a) If a strontium-90 based RTG was initially producing 128 watts, how much time would have to elapse for it to produce only 2 watts of electric power (assuming that all of the other electronic bits were still functioning)?

(b) How many decays would be occurring each second at this point?

(c) What element results from strontium-90 decay?

33-11. Fluorine-18 $\left(^{18}_{9}\text{F}\right)$ has a half-life of about 110 minutes and is formed when a target nucleus absorbs a proton, then emits a neutron. The fluorine-18 decays via positron emission. When the fluorine-18 is incorporated into a sugar-like molecule it is absorbed in parts of the body that are metabolizing sugar rapidly, such as cancerous tumors or active parts of the brain. The emitted positron collides with an electron and both are annihilated, producing two gamma rays. By seeing where the gamma rays originate (a process called positron emission tomography, or PET) doctors can localize a tumor, or they can study what parts of the brain are active in different kinds of thought processes.

(a) What is the target nucleus for the formation of fluorine-18?

(b) After a patient is injected with fluorine-18, how long will it take for the amount of fluorine-18 to decline to 1/64 of its initial value?

33-12. Potassium-40 makes up only 0.012% of all potassium atoms found in nature. It can decay by two principal modes, either beta decay, or electron capture followed by a gamma decay.
(a) What is the decay product from beta decay?
(b) What is the decay product from electron capture?

33-13. Potassium-40 has a half-life of 1.26 billion years. That means that the probability that any particular potassium-40 nucleus will decay is 1.74×10^{-17} in any given second. Only 11% of these decays will follow the electron capture path and result in a gamma ray being emitted.
(a) If a large banana has 600 mg of potassium (equivalent to about 9×10^{21} potassium atoms overall), how many gamma rays should come out of a large banana every second?
(b) If bananas had been around 2.5 billion years ago, how many gamma rays would a large one have emitted every second back then?

33-14. Iodine-131 has a half-life of 8.02 days. In the body it tends to concentrate in the thyroid. For patients who have had a cancerous thyroid gland removed it is often administered to kill off any thyroid cells that the surgery may have missed, because the beta particles that iodine-131 emits are sufficiently energetic to kill the surrounding cells.
(a) Suppose that you anticipate needing enough radioactive iodine to produce 1.0×10^{12} decays per second 16 days from now. How many decays should your sample be producing today?
(b) What element results from iodine-131 decay?

33-15. Amanda does some baking and has an interesting idea on how to tell when her cookies have passed their "sell-by" date and should be taken off of the shelf. She plans to bake her cookies using iodized salt, where some of the added iodine is radioactive iodine-133 with a half-life of 21 hours.
(a) If Amanda's cookies originally contain enough iodine-133 to each give off 200 decays per minute, how many decays per minute will she measure from a cookie that remains on the shelf $2 \frac{1}{2}$ days later?
(b) What will be the decay count after twice this time?

33-16. Strontium-90 (half-life, 29 years) was produced as part of the nuclear fallout from atomic bomb tests conducted in the 1950s.
(a) Approximately what fraction of the strontium-90 created in 1955 is still present in the environment?
(b) How much of the strontium-90 created in 1955 will remain in 2042?

33-17. The amount of carbon-14 in a once-living sample decreases with time.
(a) What fraction of the carbon-14 originally present in a sample is present after 9 half-lives have passed?
(b) What fraction remains after 12 half-lives have passed?

33-18. Carbon-14 has a half-life of 5760 years. A 1-gram sample of an ancient pine spear contains 1/8 as much carbon-14 as is present in a modern pine branch.
(a) Approximately how old is the spear?
(b) If a spear were 100,000 years old, why would carbon dating be inadequate for dating it?

33-19. A piece of flesh from a frozen mammoth discovered in Siberia was recently sampled for carbon-14. It contained one-fourth as much carbon-14 as a same-mass piece of flesh taken from a recently-deceased elephant.
(a) Approximately how long ago did the mammoth live?
(b) If instead the mammoth's flesh had only $\frac{1}{8}$ as much carbon-14 as does the flesh of a living elephant, how long ago did the mammoth die?

33-20. Suppose that a carbon-14 sensor can only measure down to 0.2% of the original amount of carbon-14 present.
(a) What is the oldest sample that can be dated by radiocarbon dating?
(b) Would radiocarbon-dating be useful for finding the age of meteorites? Explain.

33-21. A rock contains element A, which decays to element B with a half-life of 30 million years. Assume that the rock had no element B to begin with, and now it has 15 times as many B atoms as A atoms.
(a) How long ago was the rock formed?
(b) If the rock initially did contain some element B, would its age be younger than or older than the answer you came up with for part (a)? Defend your answer.

33-22. Uranium-238 with a half-life of 4.5 billion years decays through a series of steps to form lead-206. Suppose that a piece of an asteroid has a lead-206 atoms : uranium-238 atoms ratio of 1:1.
(a) Assuming that all of the lead-206 in the asteroid came from the uranium-238, how old is the asteroid?
(b) If some lead-206 was already present in the asteroid when it was formed, is the asteroid younger than or older than the answer you came up with for part (a)? Defend your answer.

Show-That Problems

33-23. Colbalt-60 undergoes beta decay.
Show that the beta decay of cobalt-60 $\left(^{60}_{27}\text{Co}\right)$ results in nickel-60 $\left(^{60}_{28}\text{Ni}\right)$.

33-24. When a target of aluminum-27 $\left(^{27}_{13}\text{Al}\right)$ is bombarded with neutrons, a target nucleus can absorb a neutron and eject an alpha particle.
Show that the target material then contains sodium-24 $\left(^{24}_{11}\text{Na}\right)$.

33-25. When a target of oxygen-16 $\left(^{16}_{8}\text{O}\right)$ is bombarded with protons, a target nucleus can absorb a proton and then eject an alpha particle.
Show that the target material then contains nitrogen-13 $\left(^{13}_{7}\text{N}\right)$.

33-26. Radon-212 $\left(^{212}_{86}\text{Rn}\right)$ is a radioactive gas with a half-life of 24 minutes.
Show that when radon-212 undergoes alpha decay, polonium-208 is formed $\left(^{208}_{84}\text{Po}\right)$.

33-27. Referring to the previous problem:
Show that if you start with 96 milligrams of radon-212 that you will have 91 mg of polonium-208 two hours later (and not 93 milligrams!).

33-28. Uranium-238 absorbs a neutron and then emits a beta particle.
Show that the result is neptunium-239.

33-29. Plutonium-239 has a half-life of approximately 24,000 years.
Show that it will take about 190,000 years for the amount of plutonium-239 to decrease to $^1/_{256}$ of its present amount.

33-30. A meteorite contains the radioactive isotope X, which undergoes beta decay into a stable isotope Y. Suppose that a recently-discovered meteorite contains 45 mg of X and 315 mg of Y.
First show that if it initially contained no Y, the meteorite originally contained 360 mg of X. Then show that the age of the meteorite is three times the half-life of X.

33-31. One form of cancer treatment is called *brachytherapy*. A pencil-lead thin titanium-encased "seed" of either iodine-125 $\left(^{125}_{53}I \right)$ with a half-life of 60 days, or palladium-103 $\left(^{103}_{46}Pd \right)$, with a half-life of 17 days, is inserted into cancerous tissue. The radioactive nuclei undergo electron capture decays, followed by the emission of an x-ray when an electron farther from the nucleus "falls" into an inner electron energy level to replace the electron that was captured. Show that the end products of the decays are tellurium-125 and rhodium-103.

33-32. Uranium-238 with a half-life of 4.5 billion years decays in a multi-step process.
Show that the following series of decays will result in the formation of lead-206.

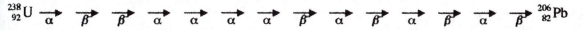

$^{238}_{92}U \xrightarrow{\alpha} \xrightarrow{\beta} \xrightarrow{\beta} \xrightarrow{\alpha} \xrightarrow{\alpha} \xrightarrow{\alpha} \xrightarrow{\alpha} \xrightarrow{\beta} \xrightarrow{\alpha} \xrightarrow{\beta} \xrightarrow{\alpha} \xrightarrow{\beta} \xrightarrow{\alpha} \xrightarrow{\beta} {}^{206}_{82}Pb$

nuclear fission, considerable energy is released when a large nucleus splits into smaller parts.
nuclear fusion, smaller nuclei combine to form a larger nucleus, and energy is released. In both
ses, the protons and neutrons (the nucleons) are more tightly bound to one another after the
action than before. As a result, as shown in the graph in Figure 34.16 on page 673 in
onceptual Physics, the mass per nucleon of atomic nuclei decreases in common fission and
sion reactions.

The difference between the mass of a nucleus and the sum total of the masses of its
nstituent nucleons is called the *mass defect*. The more tightly the nucleons are bound in the
cleus, the greater the mass defect. If in a reaction the total mass defect becomes larger, as it
es in fission and fusion, energy is released. According to Einstein's celebrated equation, the
ergy released is the *change* in the mass defect, Δm, multiplied by c^2: $E = \Delta mc^2$.

Recall from Chapters 11 and 33 that atomic masses are given in atomic mass units (amu),
here 1 amu is defined to be exactly one-twelfth the mass of a carbon-12 atom. In kilogram units
amu = 1.66×10^{-27} kg. One amu converted to energy is equivalent to 1.494×10^{-10} J = 931.5
eV (mega electron-volts, a conveniently sized unit when dealing with nuclear calculations).

The table will be useful in solving problems in this chapter. All but the first three entries give
e masses of *atoms* (including electrons), not nuclei, but changes in atomic masses are the same
changes in nuclear masses.

Particle	Mass (amu)	Particle	Mass (amu)
electron $\left(_{-1}^{0}e\right)$	0.000549	carbon-12 $\left(_{6}^{12}C\right)$	12.00000
proton $\left(_{1}^{1}p\right)$	1.007276	magnesium-24 $\left(_{12}^{24}Mg\right)$	23.985045
neutron $\left(_{0}^{1}n\right)$	1.008665	calcium-40 $\left(_{20}^{40}Ca\right)$	39.962591
hydrogen-1 $\left(_{1}^{1}H\right)$	1.007825	krypton-91 $\left(_{36}^{91}Kr\right)$	90.92344
deuterium $\left(_{1}^{2}H\right)$	2.014102	strontium-94 $\left(_{38}^{94}Sr\right)$	93.915367
tritium $\left(_{1}^{3}H\right)$	3.016049	xenon-140 $\left(_{54}^{140}Xe\right)$	139.992144
helium-3 $\left(_{2}^{3}He\right)$	3.016029	barium-142 $\left(_{56}^{142}Ba\right)$	141.916448
α-particle $\left(_{2}^{4}\alpha\right)$	4.000407	thallium-208 $\left(_{81}^{208}Tl\right)$	207.9820047
helium-4 $\left(_{2}^{4}He\right)$	4.002603	bismuth-83 $\left(_{83}^{212}Bi\right)$	211.991272
lithium-6 $\left(_{3}^{6}Li\right)$	6.015122	uranium-235 $\left(_{92}^{235}U\right)$	235.043923

Sample Problem 1
In a typical fission reaction, a slow neutron strikes a U-235 nucleus to produce two daughter nuclei and three neutrons. One such fission reaction is shown below:

$$_0^1n + _{92}^{235}U \rightarrow _{36}^{91}Kr + _{56}^{142}Ba + 3\left(_0^1n\right)$$

(a) How many joules of energy are released in this fission reaction?

Solution: We use $E = \Delta mc^2$, where Δm is the change in mass in the reaction: the initial mass of uranium and the incident neutron minus the masses of the product nuclei and the three ejected neutrons. This is equivalent to the difference between the mass of the uranium atom and that of the daughter nuclei and *two* ejected neutrons.

$$E = \left[m_{U\text{-}235} - m_{Kr\text{-}91} - m_{Ba\text{-}142} - 2m_n \right]c^2$$

$$= \left[235.043923 - 90.92344 - 141.916448 - 2(1.008665) \text{ amu} \right] \times 1.66 \times 10^{-27} \tfrac{kg}{amu} \left(3.00 \times 10^8 \tfrac{m}{s} \right)$$

$$= 2.79 \times 10^{-11} \text{ J}.$$

(b) There are 2.56×10^{21} U-235 nuclei in one gram of U-235. How many grams of U-235 would have to be fissioned to produce enough energy to heat a cup of water (250 g) from 20°C to 100°C to make tea?

Solution: $m = ?$ The energy needed to raise the temperature of the water $= c_w m_w \Delta T_w$, which is $4.18 \tfrac{J}{g \cdot C^\circ} (250g)(80^\circ C) = 83600$ J. The energy released when one gram of U-235 fissions is

$$\left(2.56 \times 10^{21} \tfrac{\text{U–235 nuclei}}{\text{gram}} \right) \left(2.79 \times 10^{-11} \tfrac{\text{Joules}}{\text{fissioning U–235 nucleus}} \right) = 7.14 \times 10^9 \tfrac{J}{g}.$$

We want to fission enough grams of U-235 so that

$$m \text{ (grams)} \times 7.14 \times 10^9 \tfrac{J}{g} = 83600 \text{ J}.$$

$$\Rightarrow m = \frac{83600 \text{ J}}{7.14 \times 10^9 \tfrac{\text{Joules}}{\text{g U–235}}} = 1.2 \times 10^{-6} g = 1.2 \text{ μg}.$$

This would be a cube of uranium 40 μm on a side. (40 μm is about half the diameter of a human hair).

Sample Problem 2

A neon-22 $\left(_{10}^{22}Ne \right)$ atom has a mass of 21.991384 amu.

(a) Calculate the mass defect of a neon-22 atom.

Solution: Neon-22 has 10 protons, 12 neutrons, and 10 electrons. Its mass defect is equal to the difference in mass between the protons and neutrons and electrons that make it up, and the mass of the final atom. That is

$$\text{Mass defect} = \left[(10m_{proton} + 12m_{neutron} + 10m_{electron}) - m_{Ne\text{-}22 \text{ atom}} \right]$$

$$= \left[10(1.007276 \text{ amu}) + 12(1.008665 \text{ amu}) + 10(0.000549 \text{ amu}) - 21.991384 \text{ amu} \right]$$

$$= 0.190846 \text{ amu}.$$

How much energy would be released if you could assemble a neon-22 atom from 10 protons, 12 neutrons, and 10 electrons?

Solution: In effect we fuse the individual nucleons to form the neon nucleus. The amount of energy is equal to the corresponding mass defect.

$$\Delta mc^2 = 0.190846 \text{ amu} \times \frac{931.5 \text{ MeV}}{\text{amu}} = 177.8 \text{ MeV}$$

$$\text{or} \quad \Delta mc^2 = 0.190846 \text{ amu} \times \frac{1.494 \times 10^{-10} \text{J}}{\text{amu}} = 2.851 \times 10^{-11} \text{ J}.$$

This may not seem like much energy, but the energy released to produce just one kilogram of neon-22 in this way is about 7.7×10^{14} J, enough to melt a cube of ice over 130 m on each side!

Example Problem 3

Bismuth-212 $\left(^{212}_{83}\text{Bi}\right)$ spontaneously undergoes alpha decay to form a thallium-208 nucleus $\left(^{208}_{81}\text{Tl}\right)$.

) Write the reaction for alpha decay of a bismuth-212.

Answer; $^{212}_{83}\text{Bi} \rightarrow {}^{4}_{2}\alpha + {}^{208}_{81}\text{Tl}$

The bismuth-212 nucleus loses 2 protons and 2 neutrons. The final nucleus has $83 - 2 = 81$ protons, and $212 - 4 = 208$ nucleons overall. This nucleus is thallium-208.

Show that the products of the decay have a smaller mass than the original bismuth atom.

Solution: The mass difference between the original nucleus and the decay products is the same as the mass difference between a bismuth-212 atom (bismuth nucleus + 83 electrons) and a helium-4 and thallium-208 atom (our decay products plus 2 and 81 electrons, respectively). Using data from the above table,

$$\Delta m = m_{\text{Bi-212}} - (m_{\text{Tl-208}} + m_{\text{He-4}}) = 211.991272 \text{ amu} - (207.9820047 + 4.002603) \text{ amu}$$

$$= 0.0066643 \text{ amu}.$$

) Is energy absorbed or released by this reaction?

Answer: Energy is released, as the total mass of the products is less than the original mass.

oblems

34-1. Amanda sees that the element lithium is to the left of iron in the periodic table, and hence to the left of the bottom of the curve of Figure 34.16 in the textbook.
   (a) Would lithium be a candidate for releasing energy by fission, or by fusion?
   (b) Calculate the mass defect for a lithium-6 $\left(^{6}_{3}\text{Li}\right)$ atom.

34-2. One of the many ways uranium-235 can fission when it is hit with a neutron is to form xenon-140 $\left(^{140}_{54}\text{Xe}\right)$ and strontium-94 $\left(^{94}_{38}\text{Sr}\right)$ as daughter nuclei.

    (a) Write this nuclear fission reaction. (Remember to include the neutron at the beginning, and the appropriate number of neutrons at the end).

    (b) Calculate the energy released in one fission reaction.

34-3. Nuclear power plants are typically about 33% efficient in converting thermal energy to electricity.

    (a) The fission of a single U-235 nucleus typically releases an average of 3.0×10^{-11} J. How much energy is released from the fission of 1.00 kg of U-235?

    (b) How many kg of U-235 are "burned" every day in a 1000 MW electric power plant?

    (c) If the power plants were more efficient, would more or less uranium fuel be needed for the same power output? Defend your answer.

34-4. One reaction scheme for running a fusion reactor involves firing a neutron at a lithium-6 nucleus to produce tritium (hydrogen-3) and then fusing the tritium to a deuterium (hydrogen-2) nucleus to form helium and another neutron:

        (1) $^{1}_{0}\text{n} + ^{6}_{3}\text{Li} \rightarrow ^{3}_{1}\text{H} + ^{4}_{2}\text{He}$

        (2) $^{3}_{1}\text{H} + ^{2}_{1}\text{H} \rightarrow ^{4}_{2}\text{He} + ^{1}_{0}\text{n}$

    (a) How much energy is required or produced in each step (use the atomic masses from the table on page 231)?

    (b) How much energy overall?

    (c) Calculate the amount of energy produced by the fission of 3.00 kg of lithium-6 and the subsequent fusion of 1.5 kg of tritium with 1.00 kg of deuterium? (This is 3.0×10^{26} atoms of each material).

34-5. Assume that the fusion reaction of the previous problem were spread out over a period of 1 day.

    (a) What would be the average thermal power output (in MW) of this reactor?

    (b) If 40% of this heat energy could be turned into electrical energy, what would be the electrical power output of this fusion reactor?

34-6. An efficient coal-burning power plant consumes about 10 tons of coal to produce a megawatt of electrical power for one day.

    (a) How many tons of coal are equivalent energy-wise to 3 kg of lithium-6 and 1 kg of deuterium? (Refer to your results from Problem 34-4.)

    (b) A railroad car can carry about 110 tons of coal. How may railroad cars of coal are required per day to produce the same power obtained from the fusion of 4 kg of fusion fuel per day?

34-7. The biggest fusion bomb ever detonated was the *Tsar Bomba*, exploded on the 10th of July, 1961, in a test by the former Soviet Union. The bomb had an explosive yield of about 50 megatons of TNT (each megaton is 4.2×10^{15} joules). The bomb itself was about 8 meters long and 2 meters in diameter. (It was a one-of-a-kind bomb rather than one of a family of bombs.)

    (a) How much mass was converted to energy in this bomb?

    (b) If an asteroid moving at 18 km/s were to have this much kinetic energy in striking the Earth, what would be the approximate mass of the asteroid?

34-8. A "kiloton" is the amount of energy released when 1000 tons of TNT explodes, about 4.2×10^{12} J. Suppose that a single uranium fission reaction releases about 3.0×10^{-11} J.

(a) Approximately how many uranium nuclei had to fission in the 14-kiloton bomb that was dropped on Hiroshima?

(b) There are about 2.6×10^{21} uranium nuclei in a gram of uranium. How many grams of uranium fissioned in the Hiroshima bomb?

(c) Uranium has a density of about 19 g/cm^3. If the fissioned uranium were to take the shape of a cube, what would be the length of its sides?

34-9. Protons and electrons within the Sun are fused in a multi-step reaction to produce a helium nucleus, a pair of neutrinos (v), and gamma radiation (γ). The overall reaction is: $4{}^1_1p + 2{}^0_{-1}e \rightarrow {}^4_2\alpha + 2{}^0_0v + 6{}^0_0\gamma$. The neutrinos escape and the gamma rays interact with matter within the Sun, their energy eventually radiating as lower-frequency light.

(a) Calculate the energy released by the overall fusion reaction. (Ignore the neutrinos and the gamma rays in the calculation because they have no appreciable mass).

(b) The Sun's luminosity is about 3.9×10^{26} watts. How much mass is being converted to energy every second?

34-10. Continuing from the previous problem, the mass of the Sun is 2.0×10^{30} kg and it is about 5 billion years old.

(a) Assuming that the Sun has had the same luminosity for its entire lifetime, about what percentage of the Sun's mass has already been converted to energy?

(b) How would you expect this loss of mass to affect Earth's orbit around the Sun?

34-11. Californium-252 has a half-life of 2.65 years. It is used industrially and medically as a fast neutron source because it will spontaneously fission in 3% of its decays.

(a) One of the fission reactions for californium-252 is
$${}^{252}_{98}Cf \rightarrow {}^{94}_{38}Sr + {}^{154}_{60}Nd + \text{ some neutrons.}$$
How many neutrons are produced in this reaction?

(b) The average fission of californium-252 produces 3.8 neutrons. A 1 µg sample of californium-252 produces 170 million neutrons per minute. How many fissions is this each minute?

34-12. Continuing with the previous problem, note that fission represents only 3% of all decays for californium-252.

(a) How many decays are there *overall* every minute?

(b) How many minutes does it take for half of all of the californium-252 atoms in a 1 µg sample to decay? (Hint: What does the half-life of a substance represent?)

(c) Use your answers to (a) and (b) to *estimate* roughly how many atoms must be in a 1 µg sample of californium-252. (Hint: You know how many fissions per minute there are initially. You can figure how many fissions per minute there are going to be one half-life later. And, you've figured out how many minutes we're looking at . . .)

Show-That Problems

34-13. The difference between the mass of a nucleus and the sum total of the masses of its constituent nucleons is called the *mass defect*.

Show that the mass defect for magnesium-24 $\left(^{24}_{12}Mg\right)$ is 0.212835 amu.

34-14. Within the Sun, three helium atoms can fuse to form carbon-12:
$$^4_2He + ^4_2He + ^4_2He \rightarrow ^{12}_6C.$$
Show that the fusion reaction releases 1.17×10^{-12} J for each carbon-12 atom formed.

34-15. One possible fusion reaction is the combination of two deuterium (hydrogen-2) nuclei to form helium-3 and a neutron: $^2_1H + ^2_1H \rightarrow ^3_2He + ^1_0n$.
Show that this reaction produces 3.27 MeV per fusion.

34-16. Burning one metric ton (1000 kg) of dry wood releases about 20 GJ (gigajoules = 10^9 J) of energy.
Show that this is equivalent to converting 0.22 milligrams of mass to energy.

34-17. Fermium-256 $\left(^{256}_{100}Fm\right)$ spontaneously fissions to produce xenon-140 $\left(^{140}_{54}Xe\right)$ and palladium-112 $\left(^{112}_{46}Pd\right)$.
Show that 4 neutrons are released in this fission.

34-18. The total U.S. consumption of electricity in 2001 was approximately 4×10^{12} kilowatt-hours, or approximately 1.4×10^{19} J.
Show that this much energy is equivalent to converting approximately 160 kg of mass to energy (roughly speaking, the mass of a small motorcycle).

34-19. A uranium-235 nucleus will fission if it gains 4.6 MeV of energy. When a uranium-235 nucleus adds a neutron, it becomes uranium-236: $^{235}_{92}U + ^1_0n \rightarrow ^{236}_{92}U$. A uranium-235 atom has a mass of 235.043923. A uranium-236 atom has a mass of 236.045562.
Show that the mass "lost" in the process of adding a neutron to uranium-235 provides more than enough energy for the nucleus to fission.

34-20. A uranium-238 nucleus will fission if it gains 5.5 MeV. When a uranium-238 nucleus adds a neutron it becomes uranium-239: $^{238}_{92}U + ^1_0n \rightarrow ^{239}_{92}U$. A uranium-238 atom has a mass of 238.050783 amu. A uranium-239 atom has a mass of 239.054288 amu.
Show that the mass "lost" in the process of adding a neutron to uranium-238 does *not* provide enough energy for the nucleus to fission. (Uranium-238 can be made to fission, but only if the neutrons are very energetic.)

35 ░ *Special Theory of Relativity*

If you look into a rocket ship whizzing past you at very high speed, you'll see that its clocks run slower. You'll also see that the rocket ship and objects inside are shortened in the direction of motion. These changes are *relativistic effects*—the consequences of high-speed motion. Passengers on the rocket ship, interestingly, see things differently. They consider *themselves* to be at rest and *you* to be moving. To them, you are the one with the slow clocks and shrunken metersticks. In the language of relativity, you don't sense relativistic effects within your own non-accelerating frame of reference (which you consider to be "at rest"). Relativistic effects are always attributed to the frame of reference of "the other guy."

One thing that you and the other guy *will* agree on is relative speed. You will also both agree on the speed of light (although you will disagree about its frequency, or color). You will measure *different* values for distances and times, but you will agree that the time and distance measurements in the "stationary" frame of reference are linked to those measurements in the "moving" frame of reference by the factor gamma ($\gamma \equiv \sqrt{1-\left(\frac{v}{c}\right)^2}$), where v is the relative speed of the two different reference frames and c is the speed of light. The relationship for time is:

$$t = \gamma\, t_o = \frac{t_o}{\sqrt{1-\left(\frac{v}{c}\right)^2}},$$

where t is the time as measured by the "stationary" observer, and t_o is the time as measured by the "moving" observer. The elapsed time t_o as experienced by the "moving" observer will be shorter.

For distance (or length) the relationship is:

$$L = L_o \sqrt{1-\left(\frac{v}{c}\right)^2}.$$

This can also be expressed, $L = \dfrac{L_o}{\gamma}$.

A "moving" observer will measure objects in his or her own frame of reference to have a "rest-length" L_o. A "stationary" observer will measure the moving object to have a length L, contracted in the direction of motion by the factor $\sqrt{1-\left(\frac{v}{c}\right)^2}$. For example, to a moving observer, the length of a meterstick aligned along the direction of motion in his or her own frame of reference is $L_0 = 1$ m. However, you, looking in from the outside, measure the length L of the meterstick to be less than 1 m.[*]

---

[*] In order to measure the length of something, you have to note the location of its two ends *at the same time*. Since the observers in the two frames don't agree on what "at the same time" means, each will claim that the other is getting a different length because he or she is measuring the location of one end of the object, letting it move, and *then* measuring the other end of it.

Sample problem 1
Your friend Albert hops into a ship and zips away at 0.87c to visit Sirius, approximately 8.7 light-years away. *†

(a) How long do you say that his trip takes?

Focus: $t_{you\ measure}$ = ?

You say that Albert has to travel 8.7 light-years at 0.87 times the speed of light, going from here to there like any other object. So

$$\text{From } d = vt \Rightarrow t_{you\ measure} = \frac{d_{you\ measure}}{v} = \frac{8.7\ ly}{0.87c} = \frac{8.7\ c \cdot year}{0.87c} = \textbf{10.0 years.}$$

(b) How long will Albert say the trip takes?

Focus: $t_{Albert\ measures}$ = ?

From your frame of reference you are the stationary observer and Albert is moving, so Albert's clocks will run slower than your clocks by the factor gamma, γ.

$$\text{From } t_{you\ measure} = \gamma t_{Albert\ measures} = \frac{t_{Albert\ measures}}{\sqrt{1-\left(\frac{v}{c}\right)^2}} \Rightarrow t_{Albert\ measures} = t_{you\ measure}\sqrt{1-\left(\frac{v}{c}\right)^2}$$

$$= 10.0\ y\sqrt{1-\left(\frac{0.87c}{c}\right)^2} = \textbf{4.9 y.}$$

(c) What explanation can Albert provide to account for the trip taking only 4.9 years?

Answer: Albert considers himself to be the stationary observer and sees the galaxy zipping past him at 0.87c. So Albert will measure the Earth-Sirius distance to be contracted by a factor $\sqrt{1-\left(\frac{v}{c}\right)^2}$ compared to the distance that you measure. You measure the distance between Earth and Sirius to be 8.7 ly. Albert's measurement is

$$L_{Albert\ measures} = L_{you\ measure}\sqrt{1-\left(\frac{v}{c}\right)^2} = 8.7\ ly\sqrt{1-\left(\frac{0.87c}{c}\right)^2} = \textbf{4.3 ly.}$$

Sirius is moving toward him at 0.87c, so Albert will say that the time for Sirius to arrive at his ship is $t_{Albert\ measures} = \frac{d_{Albert\ measures}}{v} = \frac{4.3\ ly}{0.87c} = \frac{4.3\ c \cdot year}{0.87c} = \textbf{4.9 y!}$

Notice that you and Albert will *agree* that he'll measure a trip time of 4.9 years even though you'll *disagree* on why. You'll say he measures a shorter time because his clocks are running slow. He'll say that the time is shorter because his clocks are fine but the distance is shorter. The beauty of special relativity is that it gives both of you a way to understand what the other will measure.

---

* If Albert hopped in a ship and it suddenly accelerated to 0.87c he would be flattened against the bulkhead and would be in no position to enjoy the trip. A "real" starship would have to accelerate slowly enough for Albert to withstand the forces on him. However, "thought experiments" like the one in this problem, used by Einstein himself, provide a good way to understand the meaning of relativity.

† A *light-year* (abbreviated ly) is a unit of distance, equal to how far light travels in one year at its speed c (3×10^8 m/s). If we're willing to accept answers that give time in units of years and speeds as some fraction of the speed of light (or light-years per year) we don't need to convert to the more standard meters and seconds.

Sample Problem 2
You measure your space speeder to be 36 m long, but observers on the space station you fly past measure it to be 27 m long. How fast are you moving relative to the space station?

Focus: $v = ?$

From the space station's frame of reference your moving ship appears shorter. We start out with $L_{\text{station measures}} = L_{\text{you measure}} \sqrt{1 - \left(\frac{v}{c}\right)^2}$, from which we want to get v. To isolate v we'll need to get both L's on one side of equation, square both sides to get rid of the square root sign, get $\left(\frac{v}{c}\right)^2$ by itself, then take a square root. Here we go:

$$\frac{L_{\text{station measures}}}{L_{\text{you measure}}} = \sqrt{1 - \left(\frac{v}{c}\right)^2} \Rightarrow \left(\frac{L_{\text{station measures}}}{L_{\text{you measure}}}\right)^2 = 1 - \left(\frac{v}{c}\right)^2 \Rightarrow \left(\frac{v}{c}\right)^2 = 1 - \left(\frac{L_{\text{station measures}}}{L_{\text{you measure}}}\right)^2$$

$$\Rightarrow \frac{v}{c} = \sqrt{1 - \left(\frac{L_{\text{station measures}}}{L_{\text{you measure}}}\right)^2} \Rightarrow v = \left(\sqrt{1 - \left(\frac{L_{\text{station measures}}}{L_{\text{you measure}}}\right)^2}\right) c = \sqrt{1 - \left(\frac{27 \, m}{36 \, m}\right)^2} \, c = \mathbf{0.66c.}$$

Problems

35-1. Thomas, a rhino, is 2.5 meters long when at rest.
   (a) How long will you measure him to be when he runs past you at $0.80c$?
   (b) How much time will you say it takes for him to pass you?

35-2. A ship zips by at $0.60c$. Someone aboard is making a "3-minute egg" for breakfast.
   (a) What cooking time will you measure for the egg?
   (b) Why should you not be surprised when the egg turns out to be perfectly cooked, rather than overcooked? (Hint: How much does the chef age during the time the egg is cooking?)

35-3. You measure the length of a moving meterstick to be 0.40 m.
   (a) How fast is it going? (For this and the next few problems, assume that lengths are parallel to the direction of motion.)
   (b) Using $L = L_0/\gamma$, show that when both the observer and the meterstick being observed are in the same reference frame, $v = 0$ and $L = L_0$.

35-4. Before leaving the planet Hislaurels for a starship voyage, you pack a meterstick in your luggage. After the ship has settled down to a steady speed of $0.50c$, you take the meterstick out of your bag.
   (a) How long will you measure the meterstick to be?
   (b) How long does the moving meterstick appear to an observer at rest on Hislaurels?

35-5. You are standing facing forward on the floor of your starship, which is moving at $0.80c$ relative to Earth. Before you left Earth, you measured your feet to be 25 cm long.
   (a) How long will people on Earth measure your feet to be now?
   (b) Do you need to be concerned now that the shoes that you packed for the trip will be too big?

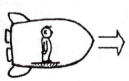

35-6. Pinocchio is concerned that Gepetto will see his long nose and realize that he has been lying. So Pinocchio decides to run past Gepetto so fast that his 10-inch long nose will appear to Gepetto to be only 2 inches long.
(a) How fast must Pinocchio run?
(b) Will Gepetto appear contracted to Pinocchio?

35-7. Lizzie is scooting down the Interstate at 17 percent of the speed of light and measures the distance between mileposts to be less than 5,280 feet.
(a) What is the distance that Lizzie measures between the mileposts?
(b) What would be the distance she measures at twice her speed?

35-8. A proton from the Sun heads toward Earth, a distance of 500 light-seconds, at 0.995c.
(a) How much time does an Earth observer measure for the proton to travel from the Sun to Earth?
(b) If the proton carried a clock, how much elapsed time would it record for the trip?
(c) What Sun-Earth distance will the proton measure (in light-seconds)?

35-9. Recent Hubble Space Telescope measurements put the distance to the open star cluster called the Pleiades (or the "Seven Sisters") to be 440 light-years.
(a) If your ship could travel there at 0.95c, would you live long enough to survive the trip?
(b) If you wanted to arrive at the Pleiades 44 years older than when you left Earth, how fast would your ship have to go?

35-10. A financial products salesman has devised a "Get-Rich-Fast" scheme. You invest $20,000 at 4% compound interest for 20 years on Earth while you are taken for a ride on a starship at 0.995c.
(a) When you return, 20 years will have passed on Earth. How much older will you be?
(b) How much money will you have in the bank when you return?[*]

35-11. You take your starship on a five-year mission to boldly collect interstellar gas samples. You leave Earth at 0.95c. After 2 ½ years have passed (according to the ship's clock) you stop and quickly turn around. You take another 2 ½ years (also on your clock) to return to Earth.
(a) What distance will you say you traveled in the first 2 ½ years? (Or, more correctly, "What length of space will you say passed you by as you remain in your starship?"
(b) How far away will you be when you temporarily stop, according to observers on Earth?
(c) How much time will have passed on Earth when you return five of your years later?
(d) The star Proxima Centauri is 4.22 light-years away from Earth. Would you be able to pass by it during your voyage?

---

[*] After one year you'd have $20,000 × $(1.04)^1$. After two years you'd have $20,000 × (1.04)^2$. After three years you'd have $20,000 × (1.04)^3$. After twenty years. . .

35-12. • A probe leaves Earth for a very long round-trip at 0.80c. When the probe returns to Earth, scientists on Earth determine that the amount of carbon-14 in the wooden parts of the probe is half as much as the amount of carbon-14 in a living tree on Earth. The half-life of carbon-14 is 5,730 years.
(a) How much time has passed for the space probe?
(b) How much time passed on Earth?
(c) According to Earth observers, how far did the space probe travel before it turned around to come back to Earth?

35-13. • One verification of special relativity theory came from measurements of muons, subatomic particles produced near the top of the atmosphere by cosmic-ray bombardment. A muon's half-life in its own frame of reference is 2.2×10^{-6} s, or 2.2 μs. Suppose that we set up two labs, one at 3300 m above sea level and the other at sea level, and that we measure 640 muons per hour at the higher-altitude lab.

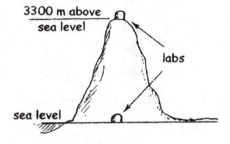

(a) Assuming that the muons are moving at a speed very close to the speed of light c (i.e., at nearly 3.00×10^8 m/s), how long will we say that it takes for them to make the trip from the higher altitude lab to the lab at sea level?
(b) How many muon half-lives is this?
(c) Ignoring special relativity effects, how many muons should remain by the time they arrive at sea level? (Remember from Chapter 33 that a half-life is the time for half of the sample of the muons to decay).
(d) Experimentally, researchers find that 320 muons per hour arrive at the lower lab. How many half-lives have passed in the muon's frame of reference?
(e) Since $t_{we\ measure} = \dfrac{t_{muon\ measures}}{\sqrt{1-\left(\frac{v}{c}\right)^2}}$, what must $\sqrt{1-\left(\frac{v}{c}\right)^2}$ be equal to?

(e) What is the speed of the muons expressed as a fraction of the speed of light?
(f) In the muon's frame of reference, they are at rest and Earth is rushing toward them. How far will they measure the distance between the upper lab and the lower lab to be?
(g) Traveling at nearly the speed of light, how much time will they measure between a point where the first lab passes them and a point where the second lab passes them?
(h) In the muon's frame of reference, if there are 640 particles when the upper lab passes by, how many should remain when they encounter the lower lab?

35-14. • You decide to visit *Wolf 359*, a star 7.8 light-years distant from Earth.
(a) If you travel there from Earth at 0.90c, how long will the folks on Earth say that it takes for your ship to arrive at *Wolf 359*?
(b) How much older will you be when you arrive than when you depart?
(c) According to you, what is the distance to *Wolf 359*?
(d) Immediately upon arrival you transmit a radio message to Earth. How long after leaving Earth will the folks on Earth hear about your arrival?
(e) You spend a year studying *Wolf 359*, send another radio message telling Earth that you are leaving, and then immediately head back home, again at 0.90c. From Earth's point of view, how long after they receive your "I'm coming home" message will you arrive home?

35-15. In 1849 Armand Fizeau measured the speed of light using a rapidly spinning toothed wheel. In the experiment, a light beam passes through a gap in the wheel, reflects from a far-away mirror, and then heads back toward the wheel. If the rotational speed of the wheel and the distance between the wheel and the mirror are just right, the reflected light will pass through the next gap in the wheel. The time it takes for the wheel to turn by one tooth would be the round-trip time for the light. If the speed of the wheel is not just right, a tooth of the wheel will block the reflected light beam.[*]

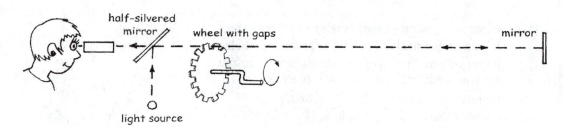

(a) Suppose that the wheel turns 150 times per second. How much time elapses during each complete rotation of the wheel?
(b) If there were 200 equally-spaced teeth around the perimeter of the wheel, how long would it take for the wheel to rotate by one tooth?
(c) If the distant mirror is located 5.0 km away from the toothed wheel, what is the round-trip distance for the light from the wheel to the mirror and back?
(d) If the reflected light nicely passes through the adjacent gap in the wheel, calculate the speed of light from the data.

_____

[*] Fizeau used a system of lenses and mirrors to focus the light. The diagram above, not surprisingly, is a simplified representation of his apparatus.

Show-That Problems

35-16. A 1.00-km long ship moves past you at 0.20c.
Show that you will measure the ship to be 980 m long.

35-17. Earth has a diameter of approximately 12,700 km.
Show that a ship passing by Earth at 0.40c will measure Earth to be about 11,600 km across.

35-18. Albert Sneedley had 15 minutes of fame on October 17, 1953.
Show that a ship passing by Earth at 0.70c would have measured his fame to last 21 minutes.

35-19. Sirius is 8.7 light-years from Earth.
Show that an astronaut making the trip at 0.50c will measure the one-way trip to take about 15 years.

35-20. Castor stays on Earth, while his twin Pollux goes for a round-trip journey at 0.80c.
Castor is 30 years older when Pollux returns to Earth.
Show that Pollux will be 12 years younger than Castor.

35-21. An electron in a particle accelerator is moving at 0.9999c.
Show that in the frame of reference of the accelerator, the electron will take 1.00×10^{-5} s to travel 3.00 km.

35-22. Refer to the 3.00-km trip taken by the electron in the previous problem.
Show that the electron experiences the 3.00-km trip as being 42.4 m long.

35-23. *Barnard's Star* is a dim red dwarf approximately 6.0 light-years from Earth.
Show that traveling there at a speed of 0.95c, you'd judge its distance from Earth to be approximately 1.9 light-years.

35-24. In your trip to Barnard's Star in the previous problem, your time of travel differs from the time that Earth observers would measure.
Show that you measure the travel time to be 2.0 years.

35-25. Marshall takes the super-train of the future from Los Angeles to Chicago, a trip of 3,600 km, at a speed of 1.8×10^8 m/s (60 percent of light speed). After a visit in hometown Chicago he takes the same super-train back to L.A. (Assume he can somehow make this round trip without being flattened by the accelerations.)
Show that Marshall experiences the trip as taking a mere 16 milliseconds to arrive in Chicago and another 16 milliseconds to return.

35-26. Referring to Marshall's round-trip voyage between Los Angeles and Chicago in the previous problem, his age at the end of the trip would be different from his age if he instead stayed at home.
Show that by taking the high-speed trip Marshall ages 8 milliseconds less than if he doesn't take the trip.